Practice for Book 9S

Contents

1 Solids *3*
2 Equivalent expressions *6*
3 Fractions *10*
4 As time goes by *13*
5 Working with rules *17*
   Mixed questions 1 *20*
6 Circumference of a circle *22*
7 Clue-sharing *25*
8 Enlargement *27*
10 Straight-line graphs *30*
11 Points, lines and arcs *34*
12 Percentage problems *35*
   Mixed questions 2 *38*

13 Ratio and proportion *40*
14 Angles of a polygon *42*
15 Using and misusing statistics *44*
16 Linear sequences *45*
17 Decimals *48*
18 Area of a circle *51*
   Mixed questions 3 *53*
19 The right connections *56*
20 Algebra problems *58*
21 Angles *62*
22 Transformations *63*
23 Trial and improvement *68*
   Mixed questions 4 *70*

1 Solids

Section A

1. Each set of cross-sections below belongs to one of these tools, but they are in the wrong order.

   For each set of cross-sections say what the tool is and give the correct order.

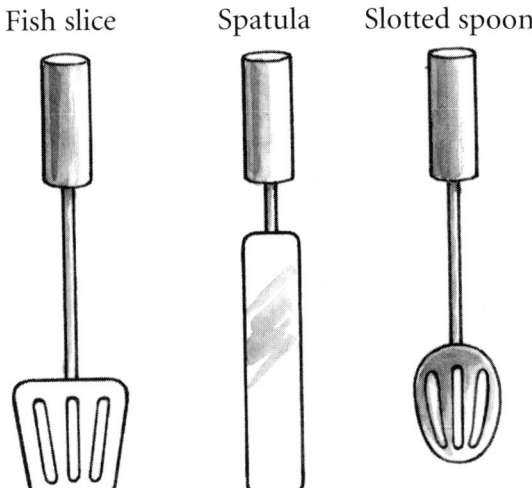

Set 1

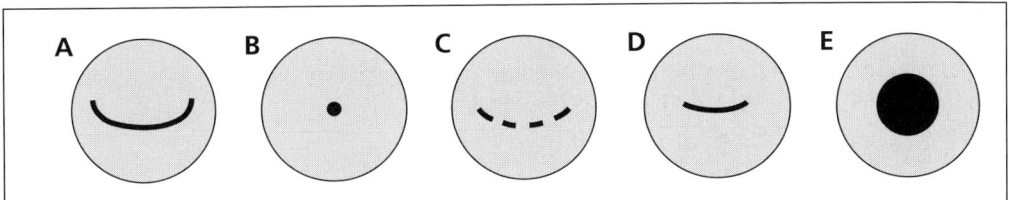

Set 2

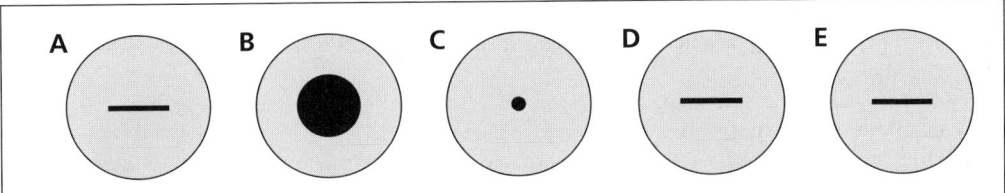

Set 3

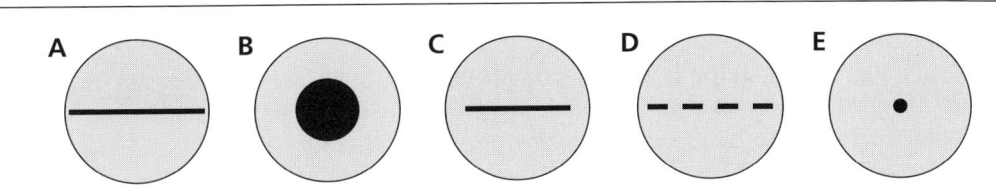

2   This key is held like this and lowered slowly into water.

   Draw five or six cross-sections for the key as it is lowered.

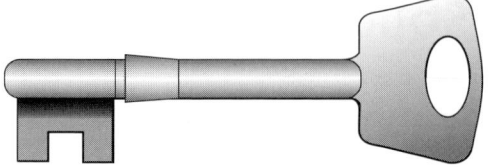

3   (a) Draw four cross-sections as this block is lowered into water like this.

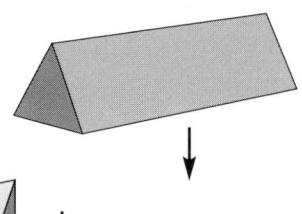

   (b) Draw some cross-sections as it is lowered like this.

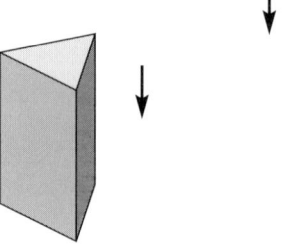

4   Find one of these objects.

   a pencil   a padlock   a safety pin

   (a) Imagine it is lowered into water.
       Draw a series of cross-sections.

   (b) Imagine it is lowered a different way round.
       Draw a series of cross-sections.

5   *Cut straight through it, any way you like, you always get a circle.*

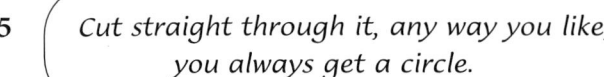

   What solid shape is she talking about?

Section B

1 Here are some pictures of prisms.
  Each prism is made from centimetre cubes or parts of them.
  Work out the volume of each prism.

(a)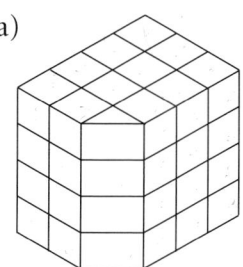

(b)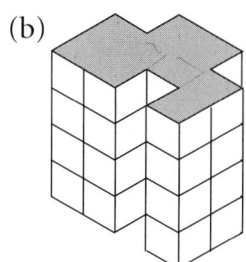

(c)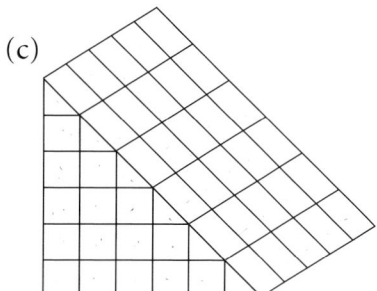

2 For each of these prisms calculate
  (i) the volume and   (ii) the surface area

(a)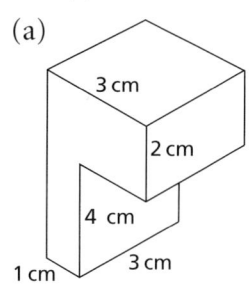

(b)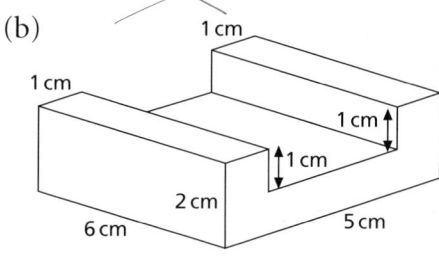

Section C

1 How many planes of symmetry do these solids have?

(a) (b)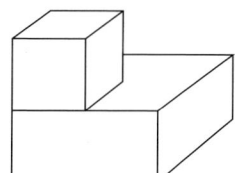

2 Equivalent expressions

Sections A and B

1. This shape is a regular hexagon.

   Which two of these expressions give the perimeter of the hexagon?

   | $6(x + 30)$ | $6x + 5$ | $6(x + 5)$ | $6x + 30$ |

   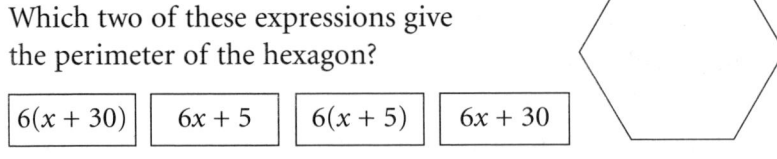

2. Multiply out the brackets in each of these.
   - (a) $3(n + 2)$
   - (b) $5(n - 3)$
   - (c) $6(n - 1)$
   - (d) $2(7 + n)$
   - (e) $2(5n + 3)$
   - (f) $3(3 + 4n)$
   - (g) $4(2n - 5)$
   - (h) $2(5n + 1)$

3. Three friends each have x seeds and each plants 2 of their seeds.

   Which two expressions below give the total number of seeds the friends have now?

   | $3x - 2$ | $3x - 6$ | $3(x - 6)$ | $2x - 3$ | $3(x - 2)$ | $2(x - 3)$ |

4. (a) For the diagram below, what is the output when the input is 4?

   Input →[− 2]→ ○ →[× 3]→ ○ →[+ 6]→ Output

   (b) Copy and complete this table for the arrow diagram above.

Input	Output
3	
5	
10	

   (c) Look at your completed table. What simple rule links the inputs and outputs?

   (d) (i) Copy and complete the arrow diagram below for an input of n. Simplify the output expression as far as you can.

   (ii) Explain how this shows the rule you found in (c) will work for any input.

5 (a) For the diagram below, what is the output when the input is 3?

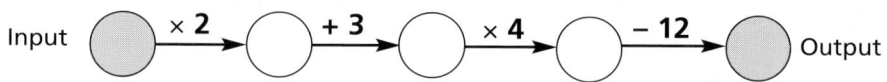

   (b) Copy and complete this table for the arrow diagram above.

   (c) Look at your completed table. What simple rule links the inputs and outputs?

Input	Output
1	
2	
10	

   (d) (i) Copy and complete the arrow diagram below for an input of n. Simplify the output expression as far as you can.

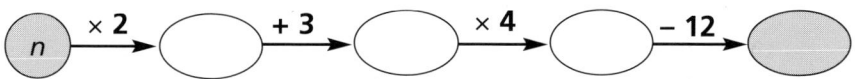

       (ii) Explain how this shows the rule you found in (c) will work for any input.

6 Multiply out the brackets in each of these.
   (a) $n(n + 5)$ (b) $m(m - 3)$ (c) $k(k - 1)$ (d) $h(4 + h)$

7 For each diagram
   (i) find the two expressions in the outer ring that **add** to give the centre expression
   (ii) find the two expressions in the outer ring that **multiply** to give the centre expression

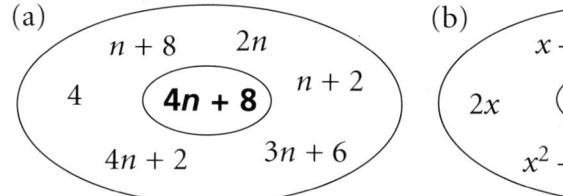

8 Multiply out the brackets in each of these.
   (a) $n(n + 4)$ (b) $2(x + 5)$ (c) $3(3y - 5)$ (d) $a(a - 6)$
   (e) $b(6 + b)$ (f) $5(2m + 1)$ (g) $p(3 - p)$ (h) $6(1 - q)$

Section C

1. Copy and complete $2n + 8 = 2(n + \blacksquare)$.

2. Factorise these expressions.
   (a) $2n + 10$
   (b) $4m - 12$
   (c) $20 + 5x$
   (d) $6y - 6$

3. Copy and complete $6a + 15 = 3(\blacksquare + 5)$.

4. Factorise these expressions.
   (a) $6p + 10$
   (b) $10a - 15$
   (c) $8b + 2$
   (d) $12q - 8$

5. (a) Factorise $10p + 25$.
   (b) Write down an expression for the length of one edge of this regular pentagon.

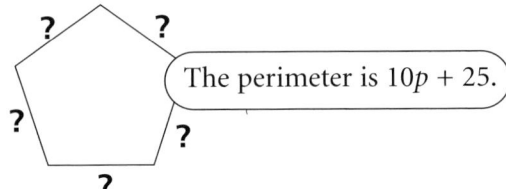

The perimeter is $10p + 25$.

6. Copy and complete $h^2 + 6h = h(\blacksquare + \blacksquare)$.

7. Factorise these expressions.
   (a) $x^2 + 7x$
   (b) $y^2 - 5y$
   (c) $p^2 + p$
   (d) $2q + q^2$

8. Factorise these expressions.
   (a) $3n + 21$
   (b) $15m - 10$
   (c) $h^2 + 4h$
   (d) $9 + 18k$
   (e) $w^2 - 6w$
   (f) $p^2 - p$
   (g) $6n - 3$
   (h) $6n + n^2$

Sections D and E

1. Simplify these expressions.
   (a) $\frac{12p}{3}$
   (b) $\frac{8q}{2}$
   (c) $\frac{15a}{3}$
   (d) $\frac{25b}{5}$

2. Which of these expressions are equivalent to $\frac{x}{3}$?

   | $\frac{2x}{6}$ | $\frac{3x}{6}$ | $\frac{4x}{12}$ | $\frac{3x}{9}$ | $\frac{5x}{20}$ |

3 Copy and complete each statement.

(a) $\dfrac{k}{4} = \dfrac{3k}{\blacksquare}$ (b) $\dfrac{n}{2} = \dfrac{\blacksquare}{10}$ (c) $\dfrac{m}{3} = \dfrac{\blacksquare}{12}$

4 Simplify these by writing each as a single fraction.

(a) $\dfrac{h}{4} + \dfrac{3}{4}$ (b) $\dfrac{a}{6} - \dfrac{5}{6}$ (c) $\dfrac{n}{5} + \dfrac{2p}{5}$

5 (a) Copy and complete $\dfrac{2}{3} = \dfrac{\blacksquare}{6}$.

(b) Simplify $\dfrac{n}{6} + \dfrac{2}{3}$.

6 (a) Copy and complete $\dfrac{b}{3} = \dfrac{\blacksquare}{6}$.

(b) Simplify $\dfrac{b}{3} + \dfrac{1}{6}$.

7 (a) Copy and complete $\dfrac{x}{5} = \dfrac{\blacksquare}{10}$.

(b) Simplify $\dfrac{x}{5} - \dfrac{y}{10}$.

8 Simplify each of these.

(a) $\dfrac{h}{8} + \dfrac{1}{2}$ (b) $\dfrac{a}{5} - \dfrac{3}{10}$ (c) $\dfrac{n}{4} + \dfrac{p}{12}$

Section F

1 In each statement, n can be any positive integer.

Which are: • always true?
• sometimes true and sometimes false?
• never true?

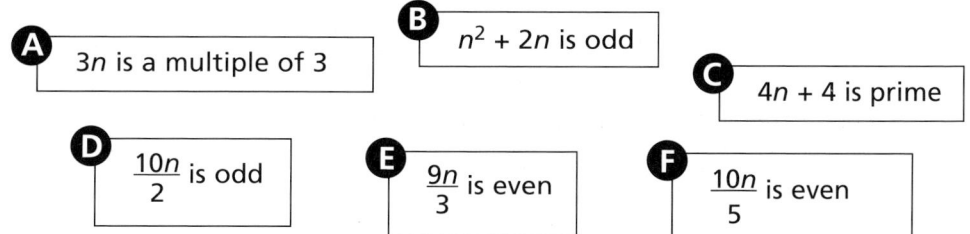

A $3n$ is a multiple of 3

B $n^2 + 2n$ is odd

C $4n + 4$ is prime

D $\dfrac{10n}{2}$ is odd

E $\dfrac{9n}{3}$ is even

F $\dfrac{10n}{5}$ is even

3 Fractions

Sections B and C

1. Jan breaks off $\frac{1}{5}$ of a large bar of chocolate.
   She eats $\frac{1}{4}$ of the $\frac{1}{5}$ that she broke off.
   What fraction of the bar of chocolate has she eaten?

2. Give answers to each of these.
   (a) $\frac{1}{4}$ of $\frac{1}{3}$
   (b) $\frac{1}{2}$ of $\frac{1}{10}$
   (c) $\frac{1}{5}$ of $\frac{1}{6}$

3. Give answers to these.
   (a) $\frac{1}{2} \times \frac{1}{6}$
   (b) $\frac{1}{8} \times \frac{1}{5}$
   (c) $\frac{1}{5} \times \frac{1}{5}$

4. Give each missing fraction.
   (a) $\frac{1}{2}$ of ? = $\frac{1}{6}$
   (b) ? of $\frac{1}{2}$ = $\frac{1}{16}$
   (c) ? of $\frac{1}{4}$ = $\frac{1}{24}$

5. One sixth of the 120 homes on a new estate are bungalows.
   (a) How many of the homes are bungalows?
   A quarter of the bungalows do not have garages.
   (b) How many of the bungalows have no garage?
   (c) What fraction of the 120 homes are bungalows with no garage?

6. Andy bakes a large slab cake and cuts it into pieces.
   He covers half the pieces with white icing.
   He puts chocolate buttons on $\frac{2}{3}$ of those
   with white icing.

   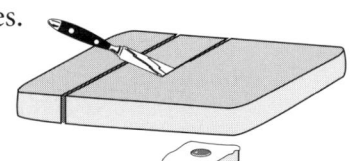

   (a) What fraction of the pieces will have white icing
       and chocolate buttons?
   (b) If he cut the original cake into 30 pieces, how many have white icing
       and chocolate buttons?

7. Work these out.
   (a) $\frac{1}{5} \times \frac{2}{3}$
   (b) $\frac{3}{4} \times \frac{3}{5}$
   (c) $\frac{3}{8} \times \frac{2}{5}$

8 Three quarters of the members of an ice-skating club are girls.
   $\frac{1}{5}$ of these girls own their ice-skates.
   What fraction of the club members are girls with their own skates?

9 Work these out, giving your answers in the simplest form. *look for cancelling*
   (a) $\frac{3}{4} \times \frac{2}{5}$ (b) $\frac{2}{3} \times \frac{3}{10}$ (c) $\frac{4}{5} \times \frac{3}{8}$
   (d) $\frac{3}{5} \times \frac{5}{8}$ (e) $\frac{2}{3} \times \frac{3}{4}$ (f) $\frac{9}{10} \times \frac{5}{6}$

10 Give each missing fraction.
   (a) $\frac{1}{2} \times ? = \frac{5}{24}$ (b) $? \times \frac{4}{5} = \frac{8}{15}$ (c) $? \times \frac{3}{5} = \frac{21}{50}$

11 A group of people tried a computer game.
   This diagram shows what they did.

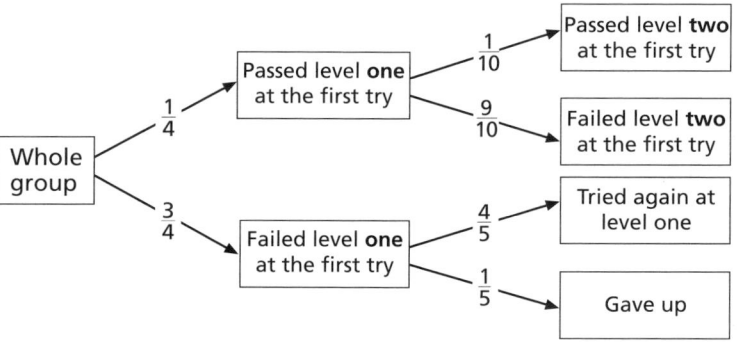

   (a) What fraction of the group
       (i)   passed both level one and level two at their first attempts
       (ii) passed level one first time but failed level two at first try
       (iii) tried again after failing level one
       (iv) failed level one at first try and gave up
   (b) Check that your answers add up to 1.

Section D

1 You can see $\frac{3}{5}$ of the CDs in this rack.
   How many CDs are there altogether?

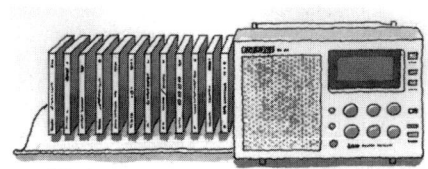

2 In a box of biscuits, 18 of the biscuits are chocolate.
   This is $\frac{3}{8}$ of the total number of biscuits.
   How many biscuits are there in the box altogether?

3 Half of the sweets in a packet are red, one third are yellow and the rest are orange.

   (a) What fraction are orange?

   (b) If there are 8 orange sweets, how many sweets are in the bag in total?

4 Bob planted $\frac{2}{3}$ of the pumpkin seeds from a packet.
   He gave away $\frac{3}{5}$ of the remainder and then had 6 seeds left.
   How many seeds were in the packet?

5 Half of the chocolates in a box are wrapped in silver foil and half in gold foil.

   (a) $\frac{1}{8}$ of the chocolates have gold foil **and** contain nuts.
       What fraction of the chocolates wrapped in gold foil contain nuts?

   (b) $\frac{1}{3}$ of the chocolates have silver foil and also have a toffee centre.
       What fraction of the chocolates wrapped in silver foil have a toffee centre?

6 Bryony made some greetings cards.
   She made some using yellow card, and some using green card.
   Some had a star design, the others had a flower design.
   This shows the fraction that she made of each type.

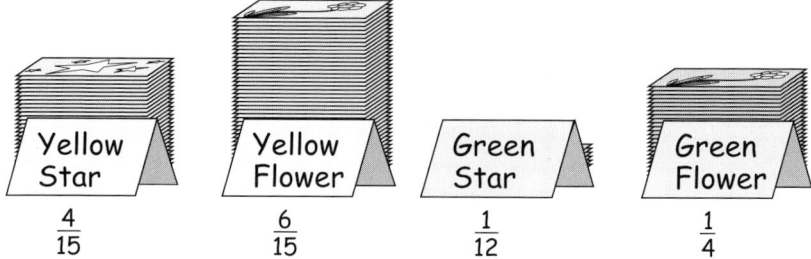

Yellow Star $\frac{4}{15}$ Yellow Flower $\frac{6}{15}$ Green Star $\frac{1}{12}$ Green Flower $\frac{1}{4}$

   (a) What fraction of the cards are yellow?

   (b) What fraction of the cards are green?

   (c) What fraction of the yellow cards have a star design?

   (d) What fraction of the green cards have a flower design?

4 As time goes by

Sections A and B

1. Water flows steadily at the same rate into these three cylindrical jars.

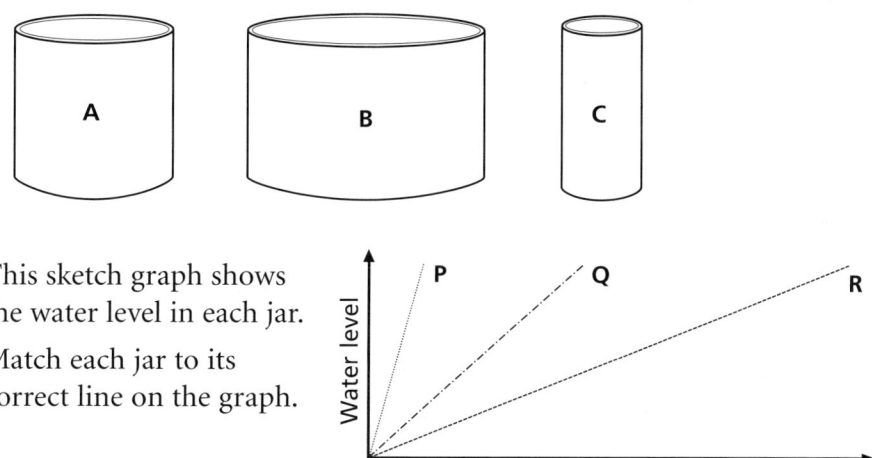

This sketch graph shows the water level in each jar.

Match each jar to its correct line on the graph.

2. Here are three containers and three graphs.
   Each container is filled steadily with water.

   Which graph goes with which container?

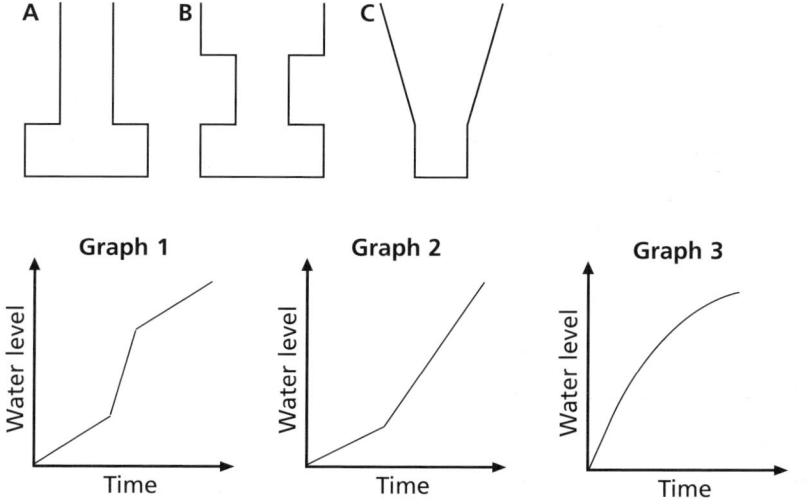

3   Imagine these containers are filled steadily with water.
    Draw sketch graphs of how the water level changes.

    (a) 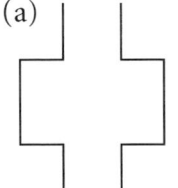   (b)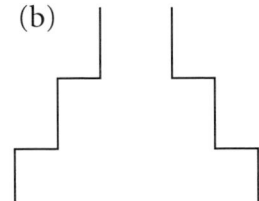

4   Rajan has a glass of cola.
    He is drinking the cola with a straw.

    He sucks the cola out of the glass at a steady rate, without stopping for breath!

    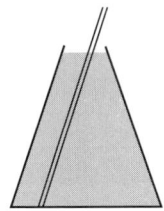

    Which graph shows the level of the cola in the glass as it empties?

    Graph 1 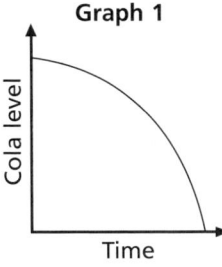   Graph 2 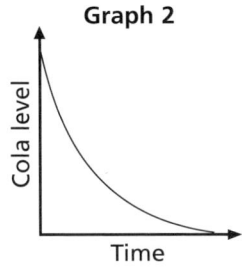   Graph 3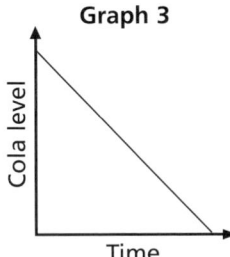

5   Imagine this container is filled steadily with oil.
    Draw a sketch graph of how the oil level changes.

    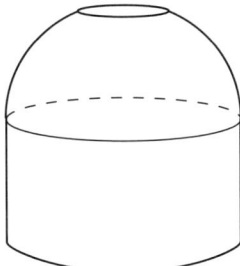

Section C

Three cars pass a tree beside a road at the same time, 12 noon.

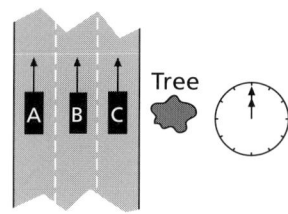

The graph shows their distances from the tree along the road between 12 noon and 2:30 p.m.

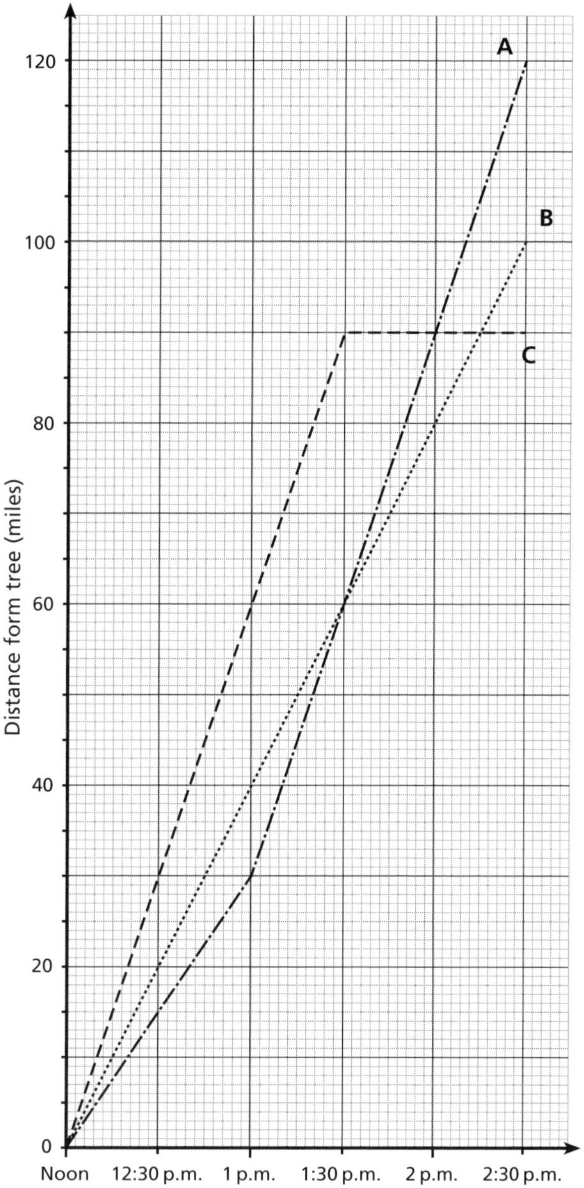

1. How far from the tree was car C at 12:30 p.m?

2. How far from the tree was each car at 1 p.m?

3. How far apart were cars A and B at 12:30 p.m?

4. When had car C travelled 75 miles from the tree?

5. At what time did car A overtake car B?

6. How far from the tree was car C at
   (a) 2 p.m.   (b) 2:30 p.m.

7. The people in car C stopped for a picnic. When did they stop?

8. What happened at 2:15 p.m?

9. What speed was car B travelling at?

Section D

1. Cass went for a cycle ride.
   This graph shows her speed.

   (a) She stopped for a few minutes.
       Between what times was this?

   (b) For 10 minutes she cycled slowly uphill.
       Between what times was this?

   (c) For 10 minutes she cycled quickly downhill.
       Between what times was this?

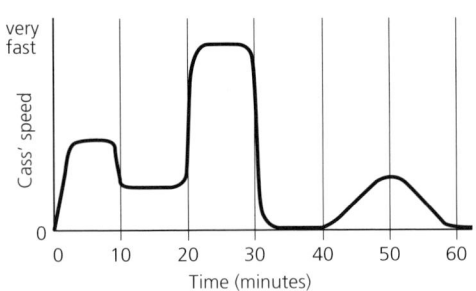

2. Jaz wrote this in his diary.

   Sketch a graph of his speed on the ride.

   Label the up-axis 'Speed' with '0' at the bottom and **very fast** at the top.

   > Went for a ride – a whole hour!
   >
   > First 10 minutes were OK – went quite fast.
   >
   > But then it took me 10 whole minutes to get up Tunnel Hill, going very slowly.
   >
   > I had to stop for 5 minutes at the top.
   >
   > But then only 5 minutes very fast down the other side.
   >
   > Met H and pushed the bike for 15 minutes, then stopped at the lake for 5 minutes.
   >
   > Then it took me 10 minutes to get home – quite fast again.

3. A man competes in a sports event.
   This graph shows his speed.

   In which of these sports do you think he was competing?

   | Sprint | Long jump | High dive | Darts |

5 Working with rules

Sections A and B

1. Solve these equations.
   (a) $3s + 2 = 26$
   (b) $5t - 4 = 36$
   (c) $19 = 2v - 8$

2. Solve these equations.
   (a) $\frac{a}{3} - 1 = 5$
   (b) $\frac{b}{9} + 4 = 5$
   (c) $\frac{c}{8} - 4 = 5$
   (d) $12 = 4 + \frac{d}{7}$

3. Solve these equations.
   (a) $4n - 3 = 15$
   (b) $3 + \frac{m}{4} = 5$
   (c) $5 + 8p = 13$
   (d) $7q - 9 = 40$
   (e) $9 = 3 + \frac{r}{10}$
   (f) $\frac{s}{8} - 3 = 2$

4. The formula for the number of chicken drumsticks Alan cooks for parties is
   $$d = 2n + 5$$
   where n is the number of guests and d is the number of drumsticks.
   (a) Use the rule to find the number of drumsticks for a party of 25 people.
   (b) If Alan cooks 37 drumsticks, how many people is he expecting at the party?
   (c) Find the value of n when $d = 85$.

5. The rule for this pattern of counters is
   $$c = 4n + 2$$
   where c is the number of counters and n is the pattern number.

   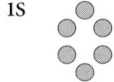 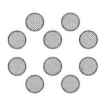 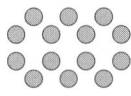
   Pattern 1   Pattern 2   Pattern 3

   (a) Use the rule to work out the number of counters in the eighth pattern.
   (b) One of the patterns uses 58 counters.
       Form an equation and solve it to find which pattern this is.
   (c) Find the value of n when $c = 150$.

Sections C and D

1. (a) Draw a flow diagram for the formula
   $T = 4N + 5$

   Your diagram should begin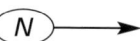

   (b) Reverse the flow diagram and find a formula for N in terms of T.
   Your formula should begin $N = \ldots$

   (c) Work out the value of N when

   (i) $T = 25$ (ii) $T = 49$ (iii) $T = 46$

2. Rearrange these to find a formula for N in terms of T each time.
   Each formula should begin $N = \ldots$

   (a) $T = 3N - 2$ (b) $T = 6N + 4$

   (c) $T = 9N - 5$ (d) $T = 8 + 2N$

3. The rule for this pattern is

   $d = 5n - 4$

   where d is the number of dots
   and n is the pattern number.

   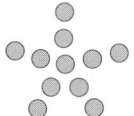

   Pattern 1   Pattern 2   Pattern 3

   (a) Work out the number of dots in pattern 9.

   (b) Find a formula for n in terms of d.

   (c) Use this new formula to find which of these patterns uses 151 dots.

4. Busby's restaurant has arranged their tables like this.

   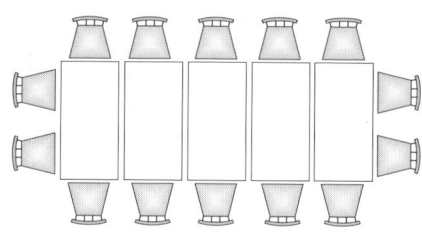

   (a) If they use 8 tables, how many chairs will they need?

   (b) Write a formula to tell you the number of chairs (c) when you know the number of tables (t).

   (c) Rearrange your formula to give a formula for t in terms of c.

   (d) Use your formula to find the number of tables needed for 42 people.

5   Rearrange each of these to find a formula for t in terms of s.
   (a) $s = \frac{t-2}{6}$
   (b) $s = \frac{t+6}{4}$
   (c) $s = \frac{3+t}{7}$

6   Rearrange each of these to find a formula for N in terms of R.
   (a) $R = 2(N+5)$
   (b) $R = 3(N-1)$
   (c) $R = 5(2+N)$

7   Rearrange each of these to find a formula for b in terms of a.
   (a) $a = \frac{b}{2} - 3$
   (b) $a = \frac{b}{5} + 4$
   (c) $a = 7 + \frac{b}{8}$

Section E

1   Rearrange each formula to make N the subject.
    Each formula should begin $N = \ldots$
   (a) $T = X + N$
   (b) $R = N - C$
   (c) $P = AN$
   (d) $Q = 4N - P$
   (e) $Y = \frac{N}{4} + X$
   (f) $Z = \frac{N-C}{2}$

2   The formula $d = 5(u + v)$ gives the distance, d metres, travelled in 10 seconds by an accelerating object if its starting speed is u m/s and finishing speed is v m/s.
   (a) Write the formula with v as the subject.
   (b) Use your formula to find the finishing speed if the starting speed was 2 m/s and the distance travelled was 40 m.

3   Rearrange each formula so that n is the subject.
   (a) $y = 5(n + x)$
   (b) $a = 3n - b$
   (c) $p = \frac{n}{q} + 8$
   (d) $c = \frac{n+s}{4}$
   (e) $c = \frac{n-3d}{e}$
   (f) $g = h + nk$

Mixed questions 1

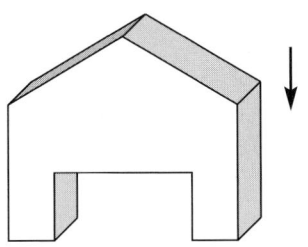

1 (a) Draw four different cross-sections as this solid is lowered into water like this.

 (b) How many planes of symmetry does this solid have?

2 Match each rectangle to the correct expression for its area.

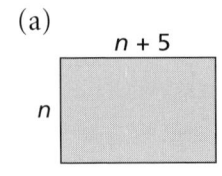 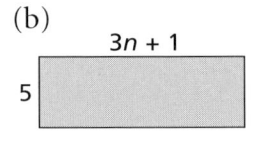 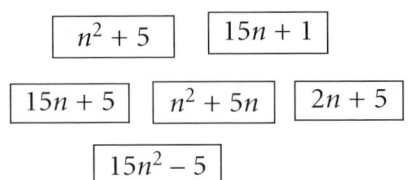

3 Calculate the volume of each of these prisms.

(a) 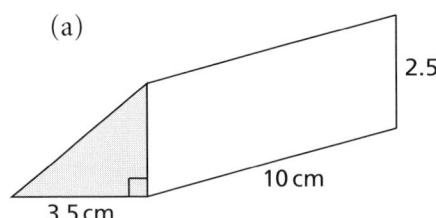   (b)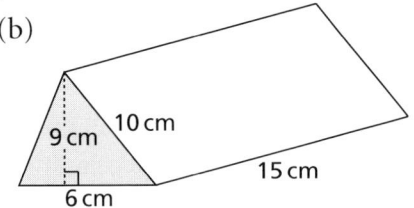

4 Five girls each have a tube of n sweets.
 Each girl eats 3 sweets.
 Which two expressions give the **total** number of sweets the girls now have?

| $5(n + 3)$ | $5n - 3$ | $5(n - 3)$ | $5n + 3$ | $5n - 15$ | $5n + 15$ |

5 Work these out, simplifying where possible.
 (a) $\frac{1}{4}$ of $\frac{1}{2}$ (b) $\frac{1}{3} \times \frac{1}{5}$ (c) $\frac{2}{3} \times \frac{1}{7}$ (d) $\frac{3}{4} \times \frac{8}{9}$

6 Factorise these expressions.
 (a) $7n + 21$ (b) $9n - 6$ (c) $n^2 - 8n$ (d) $5n + n^2$

7 What solid shape can be cut to give a circle and can also be cut to give a rectangle?

20

8 Jade wrote this is her diary.

> Went out to visit Alex. First I cycled along Holt Lane for 10 minutes going pretty steadily – then down Green Hill for 2 minutes getting faster and faster. I had built up speed coming down the hill so I went quite fast for about 15 minutes. Then I stopped for 5 minutes to watch some birds. Then I cycled steadily for 13 more minutes to reach Alex's house. The journey took $\frac{3}{4}$ of an hour in total.

Sketch a graph of her speed on the cycle journey.
(Label your speed axis **0** at the bottom and **Very fast** at the top.)

9 Find each missing fraction.
   (a) $\frac{1}{2}$ of ? = $\frac{1}{8}$ (b) $\frac{1}{3} \times $? = $\frac{5}{24}$ (c) ? $\times \frac{2}{3} = \frac{4}{9}$ (d) ? of $\frac{2}{9} = \frac{1}{9}$

10 Some solids are made by gluing 1 cm cubes together in a line.

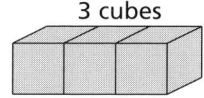
3 cubes

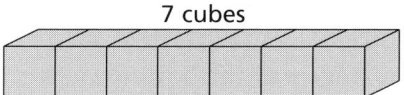
7 cubes

   (a) Calculate the surface area of each solid above.
   (b) What will be the surface area of the solid made from 10 cubes?
   (c) Write a formula that begins $s = \ldots$ that gives the surface area of a solid made with n cubes.
   (d) Find a formula for n in terms of s.
   (e) One solid has a surface area of 150 cm^2.
       How many cubes is it made from?

11 Simplify these expressions.
   (a) $\frac{8x}{4}$ (b) $\frac{4x}{12}$ (c) $\frac{a}{2} + \frac{1}{2}$ (d) $\frac{b}{3} - \frac{5}{6}$

12 Rearrange each of these to find a formula for x in terms of y.
   (a) $y = \frac{x}{3} + 5$ (b) $y = 6(x - 5)$ (c) $y = \frac{x+5}{7}$

13 $\frac{2}{3}$ of the chocolates in a box have soft centres.
    $\frac{3}{5}$ of these soft centres are fruit-flavoured.
    What fraction of the chocolates in the box are fruit-flavoured soft centres?

6 Circumference of a circle

Sections A and B

You should do all the questions in this section without a calculator.

1 Work out what the circumference is each of these tins, roughly.

   (a) Diameter 5 cm   (b) Diameter 12 cm   (c) Diameter 17 cm

2 This wedding cake has three layers.
   A ribbon goes round the outside of each layer.

   (a) The diameter of the bottom cake is 32 cm.
       Roughly how long will the ribbon be?

   (b) The diameter of the middle layer is 24 cm.
       Roughly how long will this ribbon be?

   (c) The diameter of the top cake is 16 cm.
       Roughly how long will the top ribbon be?

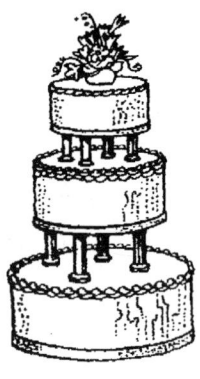

3  This groundsman is marking out the centre circle on a football pitch.
   The circle has a radius of 11 m.

   Roughly how far does he walk to mark it?

4 This traffic island is 15 metres across.
   New kerbstones are being laid around the circumference of the island.
   Each kerbstone is $\frac{1}{2}$ metre long.

   Roughly how many will be needed?

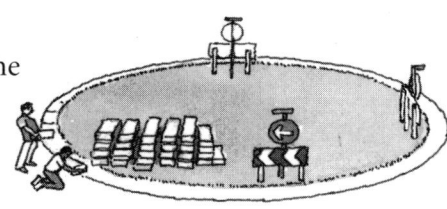

5   On old sailing ships, sailors pulled up the anchor by pushing round a windlass, like this.

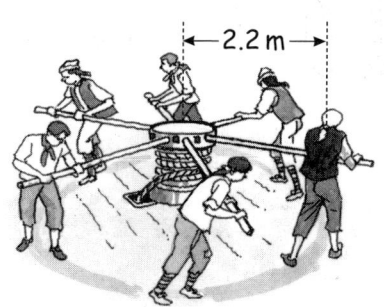

(a) Roughly how far would each sailor walk as the windlass made one turn?

(b) If it took 100 turns to bring up the anchor, how far would each sailor walk?

6   Mary and Paul are measuring some trees. They measure the circumference of each tree. Then they work out roughly the diameter and radius of each tree.

Copy and complete the table below for some trees they measured.

Type of tree	Circumference	Diameter	Radius
Oak	1.2 m		
Silver birch	0.3 m		
Horse chestnut	2.1 m		
Yew	4.1 m		
Beech	2.5 m		

Section C

Use the π key on your calculator for these questions.
Give each answer to the nearest 0.1 cm.

1   Work out the circumference of each circle.

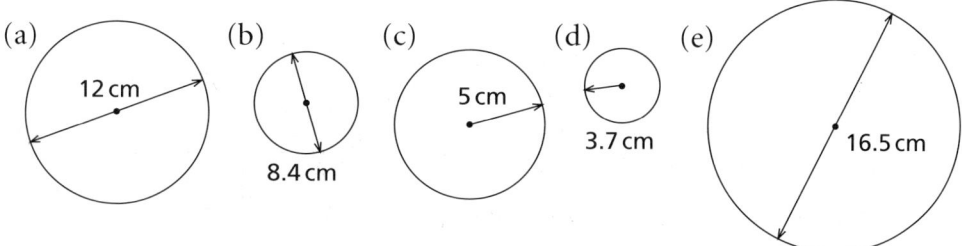

23

2. The diameter of the top of a bucket is 26.5 cm and the diameter of the bottom of the bucket is 19 cm.

How much longer is the circumference of the top than the circumference of the bottom?

3. Find the circumference of each of the following circles.
   (a) Radius 7 cm
   (b) Diameter 15 cm
   (c) Radius 15 cm
   (d) Radius 0.4 cm
   (e) Diameter 75 cm
   (f) Radius 90 cm

4. A gardener is marking out a new circular flower bed.
He uses a piece of string 2.5 metres long attached to a stake put into the middle of the flower bed.

What will be the circumference of the flower bed?

Section D

1. Rozina is designing circular bangles.
She bends silver wire to make each bangle.
   (a) What will the diameter of a bangle be if it is made from a piece of silver wire 22.5 cm long?
   (b) What will be the diameter of bangles made from wire of these lengths?
      (i) 25.3 cm   (ii) 30.8 cm   (iii) 35.6 cm

2. Calculate the radius of a circle whose circumference is
   (a) 18 cm
   (b) 55 cm
   (c) 7.6 cm
   (d) 255 cm

3. A satellite is orbiting the Moon. Each complete orbit is 11 020 km.
   (a) Assuming that the orbit is circular, how far away from the centre of the Moon is the satellite?
   (b) The radius of the Moon is approximately 1740 km.

   Roughly how far above the surface of the Moon is the satellite?

7 Clue-sharing

Sections A and B

1. List the following.

   (a) The prime numbers that are less than 20

   (b) The multiples of 6 between 10 and 40

   (c) The factors of 36

   (d) The first eight square numbers

   (e) The prime numbers between 30 and 50

2. Sue and Ravi make puzzles for this board.
   Pairs of digits make numbers.
   The digits 0 to 9 are used for each puzzle.
   A digit can only be used once in each puzzle.
   Solve each puzzle.

   Clues for Sue's puzzle
   - The two numbers in the right-hand column are multiples of 21.
   - The top left-hand number is a multiple of 5.
   - The right-hand digit of each number is prime.
   - The two numbers in the bottom row are less than 50.
   - The sum of the numbers in the top row is 158.

   Clues for Ravi's puzzle
   - The sum of the two numbers in the left-hand column is 90.
   - The numbers in the bottom row are less than 40.
   - The left-hand digit of each number is odd.
   - The numbers in the top row are greater than 60.
   - The numbers in the top row are multiples of 12.
   - The bottom right-hand number is a multiple of 10.

3 Daniel makes a puzzle for this board.

   He uses the numbers 1 to 10 for his puzzle.

   Solve this puzzle.

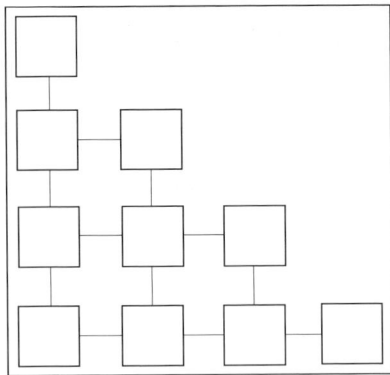

> **Clues for Daniel's puzzle**
> - The sum of the numbers in the second column is 15.
> - The numbers in the bottom row are factors of 8.
> - The numbers in the first column are odd.
> - The numbers at the corners of the triangle are square numbers.
> - The sum of the numbers in the third row down is prime.
> - The difference between the numbers in the third column is 2.
> - The numbers in the third column are factors of 24.

4 Copy and complete this crossnumber puzzle.

   Across
   1 A factor of 42
   3 A square number
   5 A square number
   6 A multiple of 8
   7 A multiple of 13
   8 A prime number that has prime numbers as digits

   Down
   2 The product of 77 and 63
   4 The sum of 697 and 698

8 Enlargement

Sections A and B

1 The triangle XYZ has been enlarged by scale factor 3 to give X'Y'Z'.

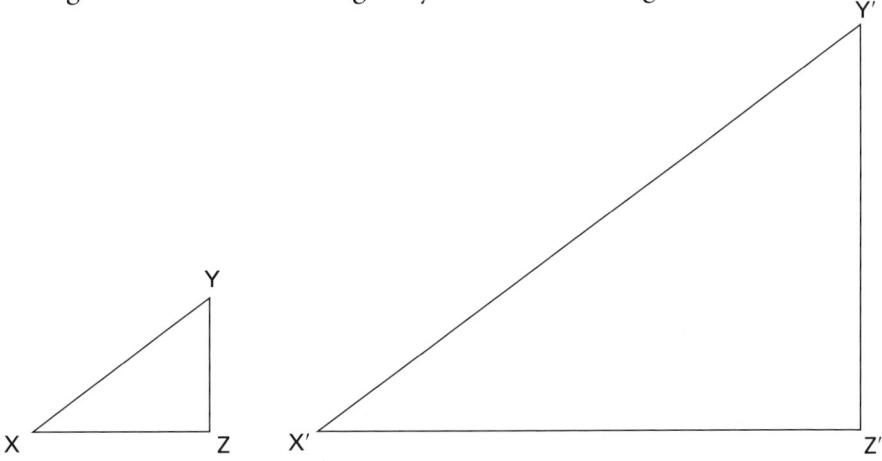

   (a) Angle YXZ = 37°.
       Write down the size of angle Y'X'Z'.

   (b) The perimeter of XYZ is 7.2 cm.
       Calculate the perimeter of X'Y'Z'.

2 (a) On squared paper draw a pair of axes, both numbered from −14 to 14.
       Plot the points (3, 7), (3, 10), (−1, 10) and (−1, 8).
       Join them up and label the shape A.

   (b) Draw the image of shape A after an enlargement of scale factor 2 with
       the point (10, 12) as the centre of enlargement.
       Label the image B.

   (c) Draw the image of shape A after an enlargement of scale factor 5 with
       the point (1, 12) as the centre of enlargement.
       Label the image C.

   (d) Shape C is the image of shape B after an enlargement.
       (i) What is the scale factor of this enlargement?
       (ii) Find the coordinates of the centre of enlargement.

3 A shape has vertices P (2, 3), Q (5, ⁻2) and R (⁻3, 1).
   It is enlarged by a scale factor of 3 with centre of enlargement (0, 0).
   Work out the coordinates of the vertices of the image.

*4 Draw a pair of axes, both numbered from 0 to 10.
   (a) Plot and join the points A (1, 1), B (3, 0) and C (4, 3).
   (b) Enlarge ABC by a scale factor 3 with centre of enlargement (1, 0).
       Label the image points A′, B′ and C′.
   (c) In a table, show the coordinates of A, B and C with their images.
   (d) Can you find a rule that links the coordinates of a point with its image after an enlargement of scale factor 3 with (1, 0) as the centre of enlargement?

Section C

1 A triangle PQR has been enlarged using the 'ray method'.
   Part of the enlargement with some lengths is shown below.

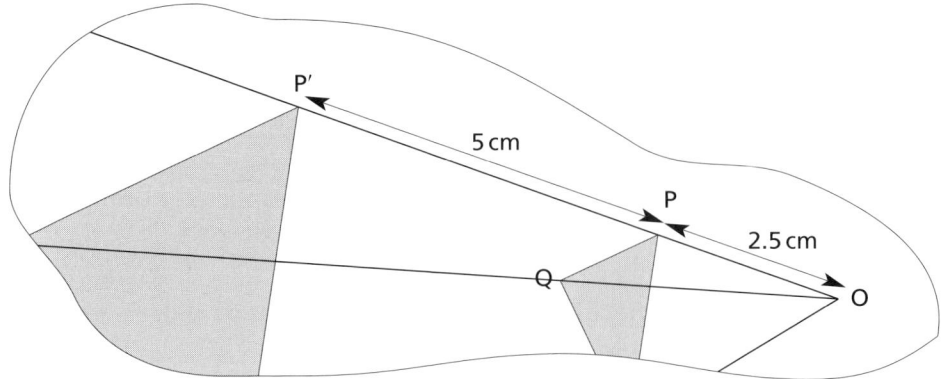

Jack has tried to work out the scale factor of the enlargement.
Work out the correct scale factor.

$\frac{5}{2.5} = 2$

so the scale factor is 2 ✗

Think again about the lengths you need, Jack.

2 In each diagram below, a triangle has been enlarged using the 'ray method'. Only part of each enlargement is shown.
Make suitable measurements to work out each scale factor.

(a)

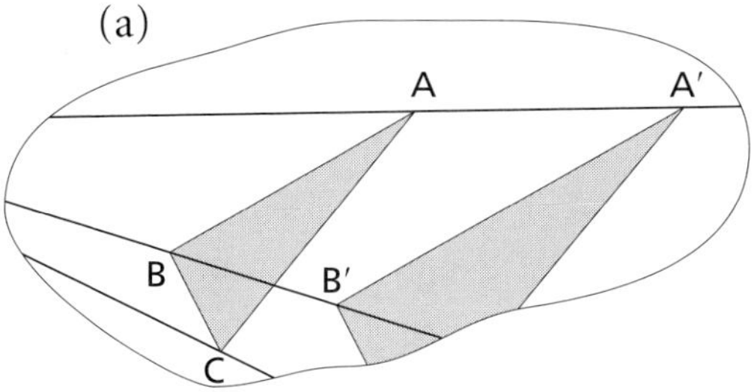

(b)

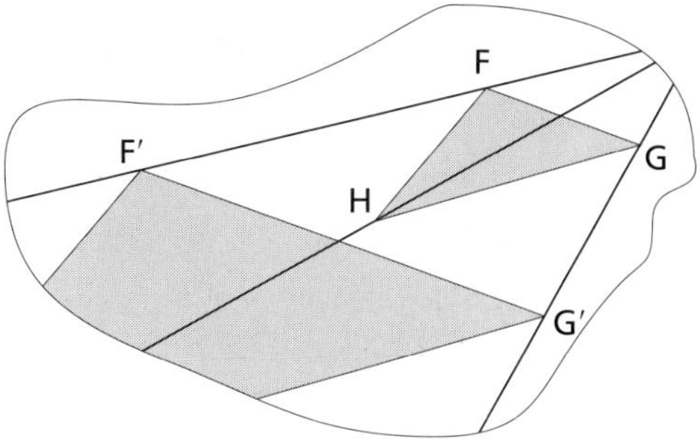

(c)

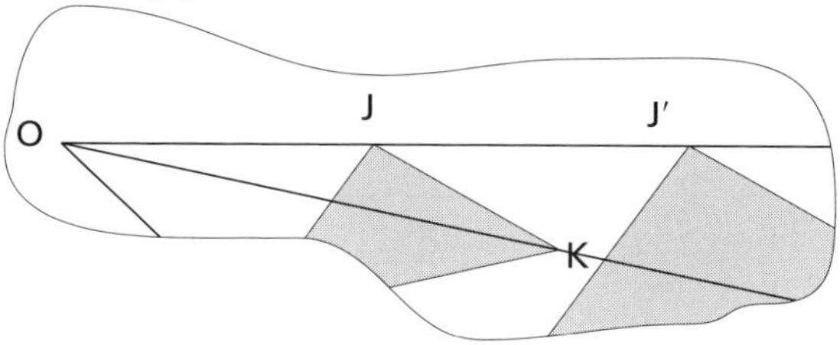

(d)

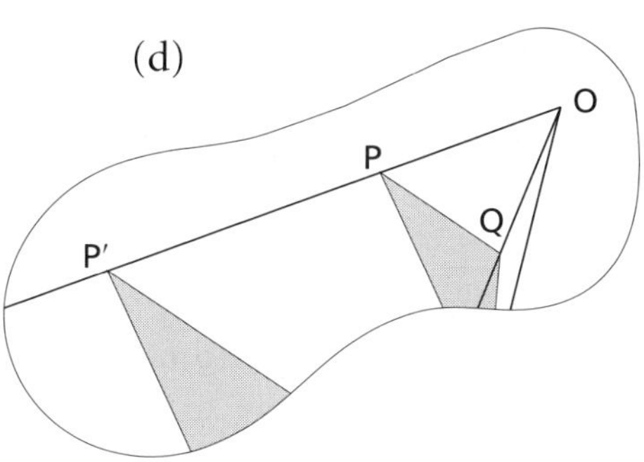

3 Shape W′X′Y′Z′ is the image of WXYZ after an enlargement with centre O and scale factor 2.

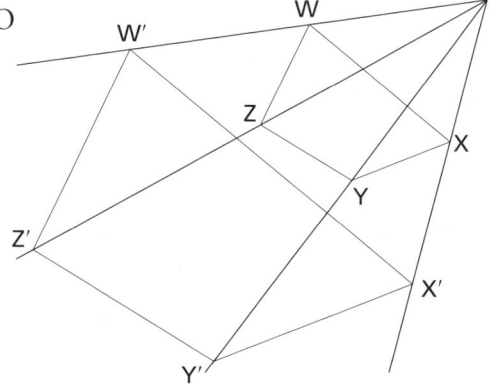

Without measuring, decide which three of these divisions are equal to 2.

$$\frac{OZ'}{OZ} \qquad \frac{WZ}{ZY} \qquad \frac{OX'}{OX} \qquad \frac{Y'Y}{OY} \qquad \frac{OW}{OW'} \qquad \frac{Z'Y'}{ZY} \qquad \frac{OZ'}{OX}$$

Straight-line graphs

Sections A and B

1 (a) Match each equation below to a table of values.

(i) $y = x - 2$

(ii) $y = 3x - 1$

(iii) $y = 2x + 2$

A

x	-1	0	1	2
y	0	2	4	6

B

x	1	2	3	4
y	-1	0	1	2

C

x	y
-2	3
0	5
2	7
4	9

D

x	y
1	2
2	5
3	8
4	11

(b) One of the tables has no matching equation.
Write down an equation that fits this table.

2 (a) Copy and complete this table for $y = 3x - 2$.

(b) Draw axes with x going from $^-2$ to 4 and y going from $^-8$ to 10.

(c) Plot the points from your table.
Check that they lie in a straight line.

(d) Draw and label the line.

x	y
-2	
0	
2	
4	

3 For each of these equations,
- find some points that fit the equation
- draw suitable axes on squared paper
- plot the points; draw and label the graph

(a) $y = x + 4$ (b) $y = 2x$ (c) $y = 2x - 3$

4 Draw a pair of axes numbered from $^-5$ to 5 along each axis.
Draw and label each of these lines on your axes.

(a) $y = 3$ (b) $y = x$ (c) $x = ^-1$ (d) $y = x - 3$

Sections C and D

1. Look at this graph.
   (a) What is the gradient of the straight line?
   (b) What is its y-intercept?
   (c) Write down its equation.

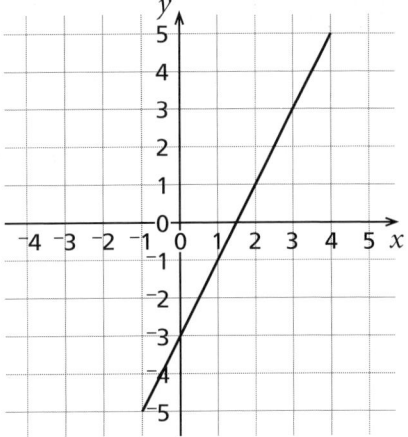

2. Look at this graph.
   (a) What is the gradient of the straight line?
   (b) What is its y-intercept?
   (c) Write down its equation.

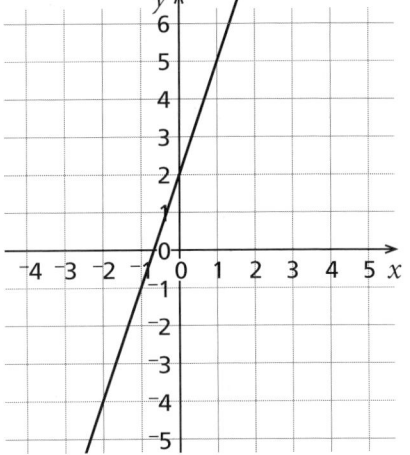

3. For each of the following lines give
   (i) the gradient and   (ii) the y-intercept
   (a) $y = 4x - 3$ (b) $y = x + 7$ (c) $y = 5x + 1$

4. (a) (i) Plot the points (1, 1) and (3, 5) on a pair of axes. Draw a line through these points.
      (ii) Find the equation of this line.
   (b) (i) Draw a line through (1, ⁻2) parallel to the first line.
      (ii) Write down the equation of this line.

Section E

1. (a) Copy and complete this table for $y = 5 - x$.
   (b) Draw suitable axes for this graph.
   (c) Plot the points you found in (a). Check that they lie in a straight line.
   (d) Draw and label the line $y = 5 - x$.

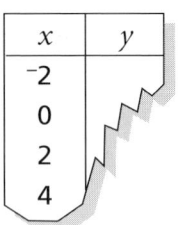

x	y
-2	
0	
2	
4	

2. For each of these equations,
   - find some points that fit the equation
   - draw suitable axes on squared paper
   - plot the points; draw and label the graph

   (a) $y = 4 - x$ (b) $y = {}^-x$ (c) $y = 6 - 4x$ (d) $y = {}^-3x + 5$

3. (a) (i) What is the gradient of line A?
       (ii) What is its y-intercept?
       (iii) Write down its equation.
   (b) Find the equation of line B.

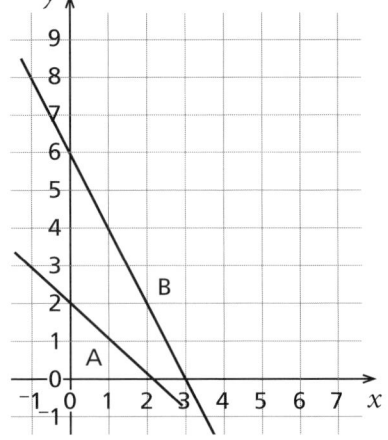

4. (a) (i) Plot the points $(0, 7)$ and $(3, 1)$ on a pair of axes. Draw a line through these points.
       (ii) Find the equation of this line.
   (b) (i) Draw a line through $(1, 1)$ parallel to the first line.
       (ii) Write down the equation of this line.

Section F

1. Draw and label the graphs of the following equations.
   (a) $x + y = 5$
   (b) $y - 2x = 6$
   (c) $y + 2x = 4$

2. A straight line has equation $y - 3x = 2$.
   (a) Rearrange the equation so that y is the subject.
   (b) What is the gradient of the line?
   (c) Write down the y-intercept of the line.

3. Find **two** pairs of parallel lines from these equations.

   A $x + y = ^-2$ **B** $y - 3x = 4$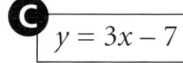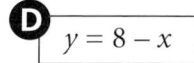

4. Match up these equations with the lines on the diagram.

   $y - 2x = ^-3$ $x + y = 3$

   $y - x = 4$ $y = 6 - 3x$

   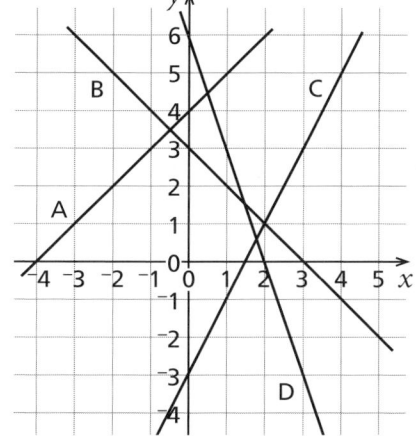

5. On squared paper draw axes with x going from $^-3$ to 5 and y from $^-2$ to 9.
   (a) On your axes draw and label the line $y = x - 1$.
   (b) On the same axes draw and label the line $y = 1 - x$.
   (c) Plot the points (4, 0) and (0, 8).
       Draw the line through the points and write down its equation.
   (d) Plot the points ($^-2$, 0) and (0, 4).
       Draw the line through the points and write down its equation.
   (e) The four lines you have drawn on your diagram form a quadrilateral.
       What special type of quadrilateral is it?

11 Points, lines and arcs

Section A

1   Construct this triangle using compasses.

    A point P inside the triangle has to obey all these rules.

    - It must be at least 6 cm from A.
    - It must be at least 3 cm from B.
    - It must be at least 5 cm from C.

    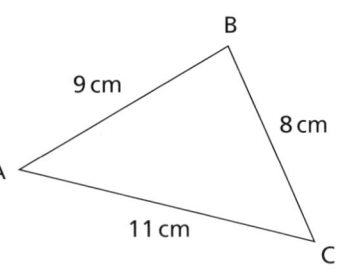

    Draw the boundary of the region where P is allowed to be. Shade the region.

2   Construct another copy of triangle ABC from question 1.
    Shade the region where point Q can go if it has to be less than 3 cm from edge AB, less than 3 cm from BC and less than 3 cm from AC.

Section B

1   Point I represents a small island in the ocean.
    Point B is a small boat 100 km south of the island.

    (a) Show these two points on a map drawn to a scale where 1 cm represents 10 km.

    (b) An exclusion zone has been declared around the island. Boats must not come within 40 km without permission.
    Draw the boundary of the exclusion zone.

    Boat B sets off on a bearing of 340°. Show this on your map.

    (c) If the boat keeps on this course, will it enter the exclusion zone? If so, for what distance will it be in the zone?

2   To be safe, the angle between a ladder and flat ground should be 70°.
    A ladder 3.5 metres long is placed at this angle against a wall.
    Use a scale drawing to find how far up the wall it reaches.

12 Percentage problems

Give all answers to the nearest 1%.

Section B

1. Pete's hourly pay increased from £6.00 to £6.90.
   What was the percentage increase in his hourly pay?

2. The price of a car increased from £10 800 to £12 100.
   What was the percentage increase in the price of the car?

3. The weekly sales of crisps at a school shop increased like this.

Ready salted	130 bags → 150 bags
Salt and vinegar	95 bags → 140 bags
Cheese and onion	215 bags → 290 bags
Barbecue	160 bags → 210 bags

   (a) Without using a calculator, which flavour do you think showed the greatest percentage increase in sales?

   (b) Work out the percentage increase for each flavour.

   (c) Comment on your answer to (a).

4. The price of a litre of petrol increased from 68.5p to 72.9p.
   What was the percentage increase in the price of petrol?

5. In 2000 the population of the United Kingdom was 59.8 million.
   In 1951 it was 50.2 million.
   What was the percentage increase over this period?

Section C

1. A bag of wet sand weighed 24 kg.
   When the sand dried out it weighed 19.6 kg.
   What was the percentage decrease in weight?

2. In a primary school, the average number of pupils in a class fell from 26.7 to 26.3. What was the percentage decrease in the average size of a class?

3. Prices were reduced in a sale.

(a) Without using a calculator, which item do you think had the greatest percentage decrease in price?

(b) Work out the percentage decrease for each item.

(c) Comment on your answer to (a).

4. In 2000, only 817 226 cases of measles were reported in the world. In 1990, 1 330 589 measles cases had been reported. What was the percentage decrease in reported cases of measles?

5. In 1980, on average, a person in Britain ate 231 grams of beef each week and 128 grams of lamb. In 2000, on average, a person in Britain ate 124 grams of beef and 55 grams of lamb.

(a) Without using a calculator, which type of meat do you think had the larger percentage decrease in the amount eaten?

(b) What was the percentage decrease in the amount of beef eaten?

(c) What was the percentage decrease in the amount of lamb eaten?

(d) Comment on your answer to (a).

Section D

1. Between January 2001 and January 2002 the number of teachers in England and Wales went up from 410 200 to 419 600. What was the percentage increase in the number of teachers?

2. In the UK in 1990, 6500 children went through the process of being adopted. In 2000, 4900 children were adopted.
What was the percentage decrease in the number of children being adopted?

3. Between 1990 and 2000 the average amount of butter eaten per person in Britain went down from 46 grams per week to 39 grams per week. At the same time the average amount of low and reduced fat spreads increased from 45 grams per week to 68 grams per week.

Calculate the percentage changes in the consumption of butter and of low and reduced fat spreads.

Section E

1. Sandra's salary of £18 500 was increased by 6%.
What was her new salary?

2. A price of £34.50 is reduced by 18%. What is the reduced price?

3. In 2001 the population of Asia was 3721 million and the population of the whole world was 6134 million.
What percentage of the population of the world lived in Asia in 2001?

4. In UK in 1995, 73 million videos were bought and 167 million were rented. In 2000, 114 million videos were bought and 186 million were rented.
   (a) Without using a calculator, do you think videos bought or videos rented increased by the greater percentage?
   (b) Calculate the percentage increase in videos bought.
   (c) Calculate the percentage increase in videos rented.

*5 The table gives the population of the world in millions.

Year	1900	1950	2001
Population (in millions)	1650	2524	6134

   (a) What was the percentage increase in the population between 1900 and 1950?
   (b) What was the percentage increase in the population between 1950 and 2001?

Mixed questions 2

1. On squared paper, draw a pair of axes both numbered from 0 to 8.
   (a) On your axes, draw and label the line $y = x + 1$.
   (b) On the same axes, draw and label the lines $x = 1$ and $y = 4$.
   (c) The three lines enclose a triangle. Label this triangle A.
   (d) Enlarge triangle A by a scale factor of 2 with (0, 0) as the centre of enlargement. Label the image B.
   (e) Write down the coordinates of triangle B.

2. Jake is designing a play area in a park. For his 'witches hat', any spectators have to be more than 5 m away from the central pole.

   He wants to cover the unsafe area with soft material and paint a yellow line round the edge.

   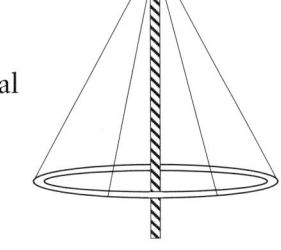

   (a) Describe the shape of the unsafe area.
   (b) What will be the length of the yellow line?

3. For each line write down
   (a) the gradient
   (b) the y-intercept
   (c) the equation

   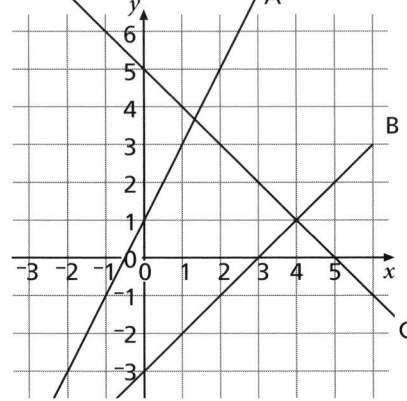

4. On squared paper, draw a pair of axes both numbered from ⁻8 to 8.
   (a) Plot, label and join the points P (1, 6), Q (7, 0) and R (1, 2).
   (b) Plot, label and join the points P′ (⁻2, 5), Q′ (7, ⁻4) and R′ (⁻2, ⁻1).
   (c) What is the scale factor of the enlargement from PQR to P′Q′R′?
   (d) What are the coordinates of the centre of enlargement?

5   A plane flies 124 km on a bearing of 127°.
    Use a scale drawing to find these distances.

    (a) How far south it has gone.   (b) How far east it has gone.

6   The price of a pair of trousers goes up from £25 to £28.
    What is the percentage increase?

7   Maia makes a puzzle for this board.
    She uses the numbers 1 to 9 for her puzzle.

    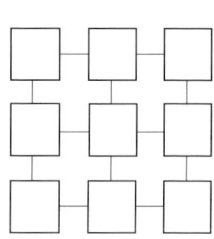

    - Numbers in the top row are prime.
    - The bottom row is made up of triangle numbers.
    - Numbers in the first column are multiples of 2.
    - Numbers in the third column are factors of 20.

    Solve the puzzle.

8   The circumference of the planet Jupiter is 445 000 km.
    What is the diameter of Jupiter, to the nearest thousand km?

9   The population of tortoises on a tropical island was 450 a year ago
    and is 415 now. To the nearest 1%, what is the percentage decrease
    in the population of tortoises?

10 (a) Make an accurate drawing of
        the triangle in this sketch.
        Draw it near the top of
        a piece of plain paper.

    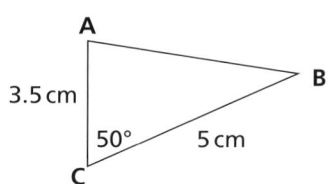

    (b) Find the perimeter of triangle ABC.

    (c) Mark a point O about 2 cm above point A.

        (i) Enlarge triangle ABC by a scale factor of 3 with O as the centre of
            enlargement. Label the image A′B′C′.

        (ii) Calculate the perimeter of triangle A′B′C′.

        (iii) What is the size of angle A′C′B′?

11 Find the gradient of the line with equation $y + 5x = 8$.

13 Ratio and proportion

Sections A and B

1 Calculate the ratio $\dfrac{\text{length}}{\text{width}}$ for each of these swimming pools.

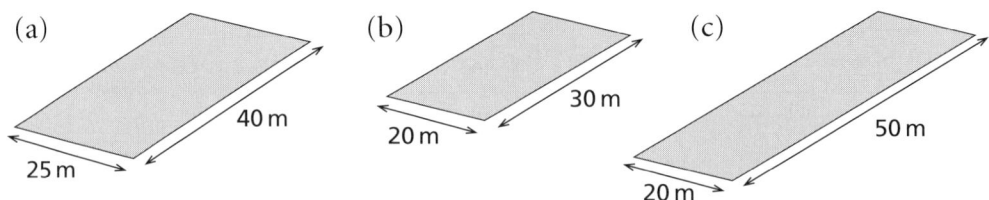

(a) 40 m, 25 m (b) 30 m, 20 m (c) 50 m, 20 m

2 The hog-nosed bat is the smallest mammal in the world.
Its average length is 3 cm and its average wingspan is 14 cm.

Calculate the ratio $\dfrac{\text{wingspan}}{\text{length}}$ for the hog-nosed bat correct to 1 d.p.

3 Here are some recipes for tropical fruit drinks.

Tropical tango	Tropical sizzler	Tropical refresher
Mix pineapple juice and mango juice in the ratio 8 : 5	Mix pineapple juice and mango juice in the ratio 5 : 3	Mix pineapple juice and mango juice in the ratio 10 : 7

(a) For each recipe, work out the ratio $\dfrac{\text{amount of pineapple juice}}{\text{amount of mango juice}}$.

(b) Which recipe will give the strongest pineapple flavour?

4 This table shows the amount of water (*W* ml) needed to make bread using different weights of wholemeal flour (*F* grams).

F	400	500	600
W	300	360	420

(a) Work out the ratio $\dfrac{W}{F}$ for each pair of values.

(b) Explain how your ratios show that *W* is not directly proportional to *F*.

5 This table shows the number of bottles (*n*) a machine fills when run for certain lengths of time (*t* minutes).

t	3	5	9	12
n	45	75	135	180

(a) Work out the ratio $\frac{n}{t}$ for each pair of values.

(b) Explain how your ratios show that *n* is directly proportional to *t*.

(c) Find an equation connecting *n* and *t*.

(d) Use your equation to find the number of bottles that would be filled in 16 minutes by this machine.

6 For the values in each table *y* is directly proportional to *x*. For each table
(a) find the missing value
(b) write down an equation connecting *x* and *y*

A
x	1.8	2.4	2.9
y	5.4		8.7

B
x	0.8	1.4	2.2
y	5.2	9.1	

C
x	20		50
y	24	42	60

D
x	5.2	6.8	
y	1.3	1.7	2.3

Section C

1 For this table, *y* is directly proportional to *x*.

x	4.4	6.7
y	7.48	

(a) Let the hidden value in the table be *n*. Use ratios to form an equation for *n*.

(b) Solve the equation to find the value of *n*.

2 To make a type of brass, 6.8 grams of zinc is mixed with 23.8 grams of copper.

How many grams of copper would be needed to mix with 11.4 grams of zinc to make the same type of brass?

 # Angles of a polygon

Section A

1. For each of these polygons,
   (a) count the number of sides
   (b) say what the sum of the interior angles will be
   (c) work out the missing angle
   (Angles with the same letter are the same size.)

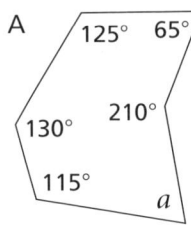 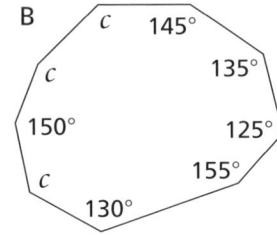 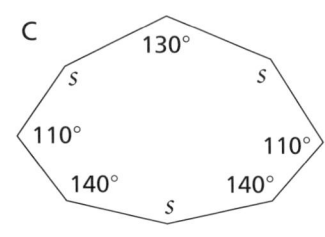

2. Work out the missing angles.

   (a) 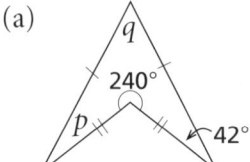   (b)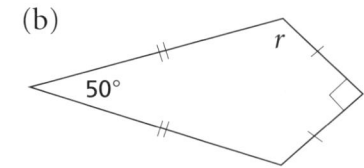

3. This hexagon has two lines of symmetry. Calculate the size of angles *a*, *b*, *c* and *d*.

   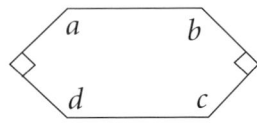

4. This octagon has rotational symmetry of order 4. The angles at the 'points' are all 40°. Find the size of the remaining four interior angles.

   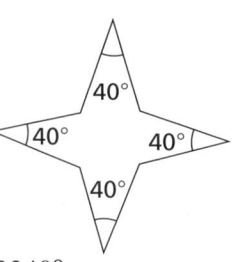

5. The sum of the angles of a polygon is 3240°. How many sides does the polygon have?

Sections B and C

1. Work out the angles that the letters stand for.

   (a) (b) (c)

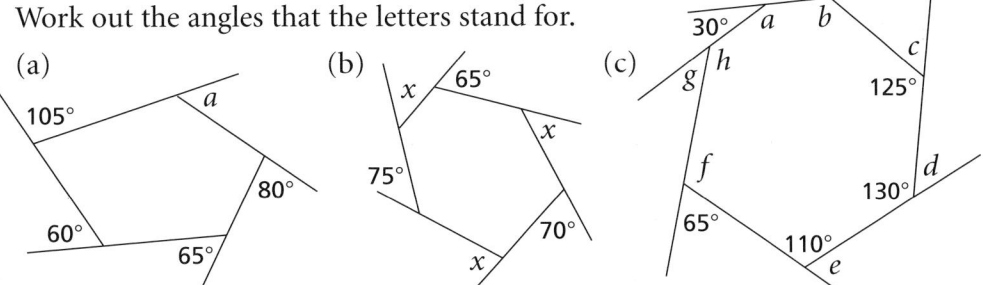

2. These are parts of regular polygons. How many sides does each one have?

   (a) (b) (c)

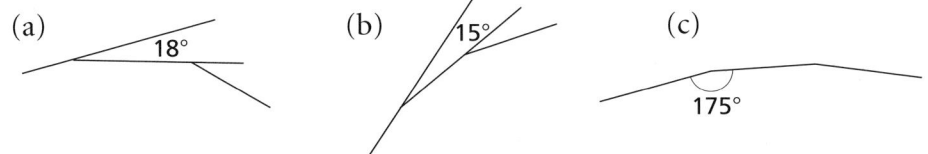

3. Draw a regular nonagon (9-sided polygon) like this.
   (a) Draw a circle.
   (b) Mark off appropriate points around the circle, using your angle measurer to help you make sure that the points are evenly spaced.
   (c) Join up these points to make the nonagon.

4. Each interior angle of a regular polygon is 168°.
   (a) What is each exterior angle?
   (b) How many sides does this regular polygon have?

5. The diagram shows a regular pentagon inside a regular hexagon, with AB being a side of each polygon.
   (a) What is the size of each interior angle of a regular hexagon?
   (b) What is the size of each interior angle of a regular pentagon?
   (c) Work out the size of angle x.

15 Using and misusing statistics

Sections A and B

1. When is a pie chart a good way to display data?
   (Give an example of a statistical investigation when you might use one effectively.)

2. From 2001 to 2002, the average attendance at Holby United games went up from 8500 to 10 000.

   (a) Which of these charts shows this information in the fairest way?

   (b) Criticise the other two charts.

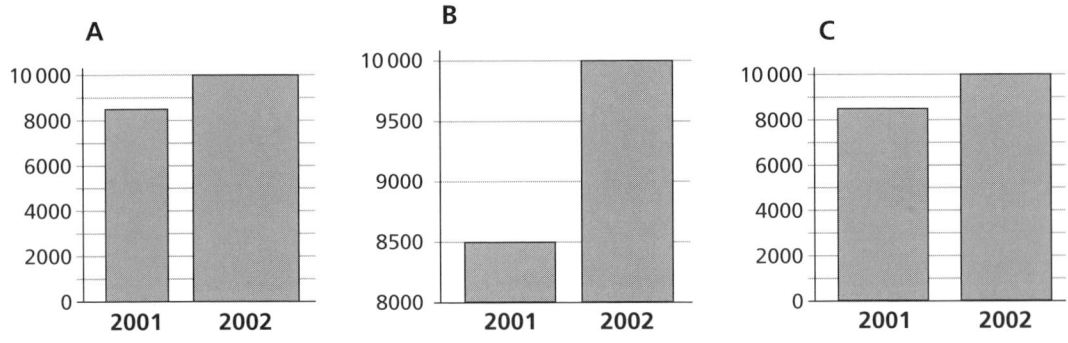

3. This chart shows the amount of sugar produced by the world's five leading sugar producing countries in 1983.

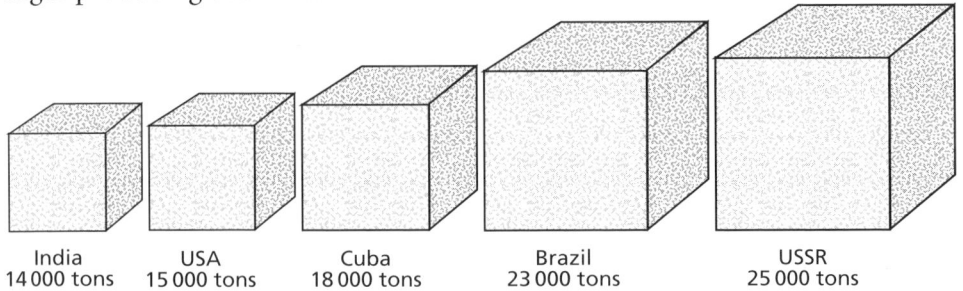

Describe what is misleading about this chart.

16 Linear sequences

Section A

1   A sequence begins 4, 9, 14, 19, 24, …
   (a) Describe a rule to go from one term to the next.
   (b) What is the 10th term of this sequence?

2   For each of the following sequences,
   • describe a rule to go from one term to the next
   • find the 10th term
   (a) 4, 7, 10, 13, 16, …
   (b) 20, 19, 18, 17, 16, …
   (c) 2, $4\frac{1}{2}$, 7, $9\frac{1}{2}$, 12, …
   (d) 3, 9, 27, 81, 243, …

3   Which of these sequences are linear?
   A 1, 3, 6, 10, 15, …
   B 5, 8, 11, 14, 17, …
   C 24, 20, 16, 12, 8, …
   D 7, 11, 15, 19, 23, …

4   Copy and complete each of these linear sequences.
   (a) 3, 9, __, 21, __, __
   (b) 5, __, __, 20, 25, __
   (c) 4, __, __, 22, 28
   (d) __, 5, __, __, 11, __

Section B

1   Each of these expressions gives the nth term of a sequence.

   A $5n$ **B** $3n + 5$ **C** $4n - 3$ **D** $n^2 + 3$
   E $200 - 2n$

   For each expression
   (a) find the first six terms of the sequence
   (b) decide if the sequence is linear
   (c) work out the 100th term of the sequence

2   Match each sequence to a correct expression for its *n*th term.
   (a) 8, 9, 10, 11, 12, …
   (b) 1, 3, 5, 7, 9, …
   (c) 7, 10, 13, 16, 19, …
   (d) 5, 11, 17, 23, 29, …
   (e) 7, 11, 15, 19, 23, …

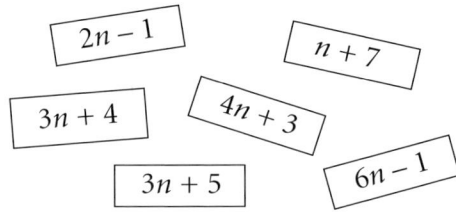

3   Match each sequence to a correct expression for its *n*th term.
   (a) 11, 10, 9, 8, 7, …
   (b) 7, 6, 5, 4, 3, …
   (c) −2, 0, 2, 4, …
   (d) −2, −7, −12, −17, …

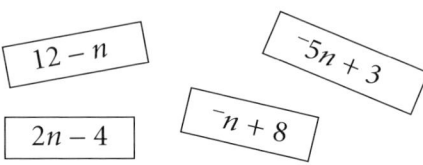

Section C

1   For each of the following linear sequences,
   • find an expression for the *n*th term
   • calculate the 30th term
   (a) 6, 10, 14, 18, …   (b) 10, 13, 16, 19, …   (c) 2, 8, 14, 20, …
   (d) 8, 9, 10, 11, …   (e) 1, 8, 15, 22, …

2   This sequence of fishes continues to the right.

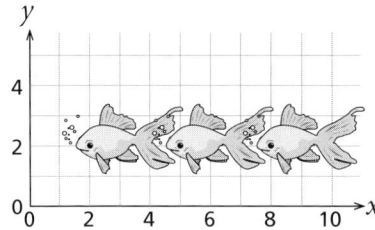

   (a) What are the coordinates of the 1st fish's eye?
   (b) Work out coordinates for
      (i) the 4th fish's eye
      (ii) the 10th fish's eye
   (c) Copy and complete this table for 10 fishes.
   (d) Give the coordinates of the *n*th fish's eye.
   (e) What are the coordinates of the 25th fish's eye?

Fish (*n*)	x	y
1		
2		
3		
4		

Section D

1. For each of the following linear sequences,
   - find an expression for the nth term
   - calculate the 25th term of the sequence

   (a) 93, 90, 87, 84, …
   (b) 157, 152, 147, 142, …
   (c) 69, 68, 67, 66, …
   (d) 82, 79, 76, 73, …
   (e) 43, 41, 39, 37, …

2. This sequence of fish continues downwards.

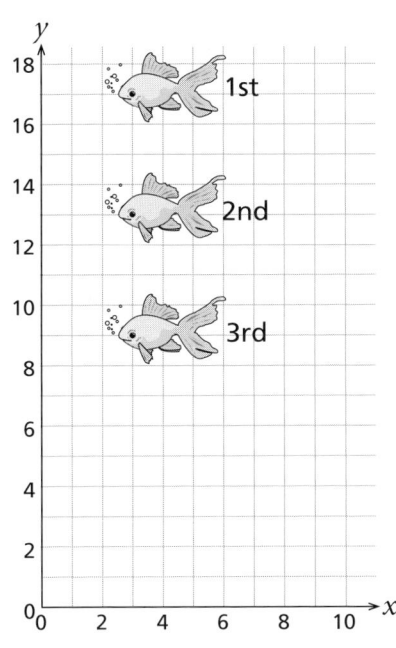

   (a) What are the coordinates of the first fish's eye?

   (b) Give the coordinates for the 5th fish's eye.

   (c) Copy and complete this table for 10 fish.

Fish (n)	x	y
1		
2		
3		
4		
5		

   (d) Give the coordinates of the nth fish's eye.

   (e) Give the coordinates of the 25th fish's eye.

17 Decimals

Do not use a calculator for any of these questions.

Section A

1 Work these out in your head.

   (a) 3.5 + 9.8   (b) 14.3 – 6.7   (c) 6 – 1.85
   (d) 3.7 + 9.85   (e) 11.3 – 4.8   (f) 9.4 – 6.05

2 These are the weights of four bags of potatoes.

   2.61 kg   2.16 kg   2.6 kg   2.06 kg

   Work out the following in your head.
   (a) The sum of the weights of the heaviest two bags.
   (b) The difference between the weights of the lightest two bags.
   (c) The difference between the weights of the heaviest and lightest bags.
   (d) The total weight of all four bags.

3 (a) In a race Bert took 28.32 seconds for the first lap and 27.59 seconds for the second lap. How long did he take in total for both laps?
   (b) The winning time for the two lap race was 54.76 seconds. How much faster was this than Bert's time?

Section B

1 Work these out. Then use the letters of the largest and smallest answers in each group of calculations to form a 6-letter mathematical word.

   E: 1.5 × 8
   F: 0.3 × 6
   R: 2.4 × 7
   S: 4.03 × 4

   C: 1.07 × 16
   N: 1.74 × 13
   P: 2.45 × 9
   T: 3.4 × 7

   A: 0.4 × 0.6
   E: 0.4 × 0.25
   M: 0.01 × 23
   O: 0.02 × 0.9

2 Do the same as question 1 with these.

P: $\sqrt{0.25}$
R: 0.03^3
T: 0.1^3
U: 0.9^2

A: 480 ÷ 100
C: 0.002 × 100
E: 0.2 ÷ 10
I: 0.04 × 100

L: 0.72 × 6.2
O: 3.2 × 1.4
Q: 0.36 × 41
S: 2.1^2

3 Copy and complete these multiplication squares.

(a)
×	0.1	0.15	0.04
4			
			0.008
0.5			

(b)
×			0.3	
5	1			
			0.12	
		1.4		0.07

4 Write these in metres.
(a) 70 cm   (b) 260 mm   (c) 75 mm   (d) 409 cm   (e) 0.65 km

5 Write these in kilograms.
(a) 450 g   (b) 2800 g   (c) 38 g   (d) 2 kg 75 grams

Section C

1 Work these out. Then use the letters of the largest and smallest answers in each group of calculations to form an 8-letter mathematical word.

A: $\frac{8.75}{5}$
D: $\frac{2.4}{8}$
I: $\frac{0.12}{4}$
P: $\frac{8.5}{5}$

E: $\frac{15}{0.5}$
O: $\frac{0.28}{0.4}$
R: $\frac{9}{0.2}$
W: $\frac{4.2}{0.6}$

F: 5.24 ÷ 0.4
M: 18.4 ÷ 0.8
T: 27 ÷ 0.06
U: 6.9 ÷ 0.03

C: 2.52 ÷ 1.2
H: 1.8 ÷ 0.15
N: 4.25 ÷ 0.25
X: 19.2 ÷ 2.4

2 Work these out, giving your answers correct to 1 d.p.
(a) 25.3 ÷ 2.4   (b) 58 ÷ 1.7   (c) 7.73 ÷ 0.9

3 (a) How many glasses each holding 0.35 litres can be filled from a container holding 5.5 litres?

   (b) An athlete ran 27 m in 2.4 seconds.
       What was his average speed in metres per second?

Section D

1 What number is half way between 7.9 and 11.45?

2 Some pupils were measuring the heights of their maize plants.
   (a) Which one is tallest?
   (b) Which one is shortest?
   (c) What is the difference in height between the tallest and shortest?

   2065 mm   210 cm   2.05 m   2 m 30 cm   209.2 cm   2.048 m   2 m 7 cm

3 Copy these and fill in the missing numbers.
   (a) 3 × ? + 0.1 = 1.3
   (b) 0.4 + ? × 0.2 = 1
   (c) (? + 0.35) × 0.1 = 0.21
   (d) ? ÷ 4 + 0.2 = 1
   (e) 0.5 − 0.7 × ? = 0.01
   (f) 0.9 × (4.3 − ?) = 0.63

4 Five people weighed a spoonful of sugar. Here are their results.

   15.2 g   14.35 g   14.7 g   15.04 g   14.65 g

   (a) What is the range of these masses?
   (b) What is the mean mass?

5 The diagram shows a packing case in the shape of a cuboid.
   (a) Find its volume in cubic metres.
   (b) Find its surface area in square metres.
   (c) How many layers of these packing cases will there be in a pile 2.4 metres high?

   0.2 m   0.7 m   1.2 m

6 The imperial unit of one stone is equivalent to 6.4 kg.
   (a) James weighs $9\frac{1}{2}$ stone. What is his weight in kg?
   (b) James' father weighs 82 kg. What is his weight in stones, to 1 d.p?

18 Area of a circle

Section A

1 Calculate, to the nearest cm², the area of a circle with radius
  (a) 6.7 cm
  (b) 4.4 cm
  (c) 7.9 cm

2 Calculate the area of these.
  (a) The outer circle of radius 8.4 cm
  (b) The inner circle of radius 5.6 cm
  (c) The shaded space between the two circles

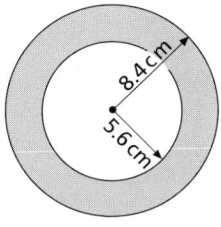

3 A circular pond is set into a rectangular lawn.

Calculate the area of grass that remains (shaded).

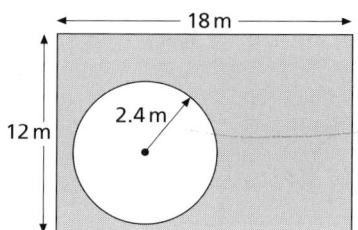

4 Calculate, to the nearest cm², the area of a circle with
  (a) diameter 12.4 cm
  (b) diameter 8.5 cm

Section B

1 Calculate, to one decimal place,
  (a) the circumference of a circle of diameter 4.2 cm
  (b) the area of a circle of radius 5.4 cm
  (c) the circumference of a circle of radius 8.8 cm
  (d) the area of a circle of diameter 12.5 cm

Section C

1. What is the radius of a circle with area 30 cm² (to the nearest 0.1 cm)?

2. Find the radius of the circle with each area.
   (a) 45 cm²   (b) 86 cm²   (c) 6.4 cm²
   (d) 34 cm²   (e) 130 cm²   (f) 18 cm²

3. The area of a circle is 43.2 cm².
   Calculate, to one decimal place,
   (a) the diameter   (b) the circumference

Section D

1. For each of these shapes calculate, to one decimal place
   (i) the perimeter   (ii) the area

   (a)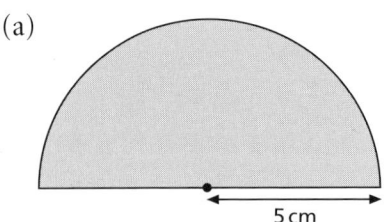

   (b)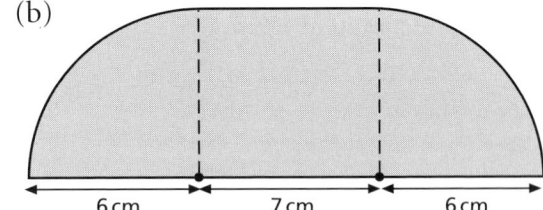

   (c)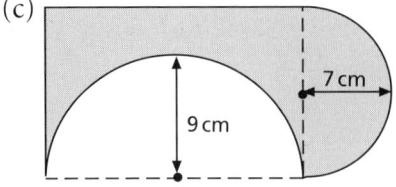

   (d)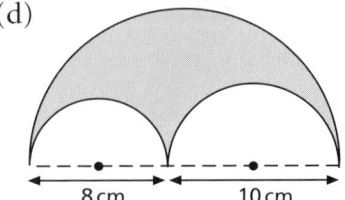

Mixed questions 3

1. The volume (V cm³) of some spheres with various radii (r cm) are shown in the table.
The volume is rounded to 2 d.p.

r	1	2	3	4
V	3.14	25.13	84.82	201.06

   (a) Work out the ratio $\frac{V}{r}$ (correct to 1 d.p.) for each pair of values.

   (b) Explain how your ratios show that V is not directly proportional to r.

2. Find the perimeter and area of the semicircle, correct to 2 d.p.

   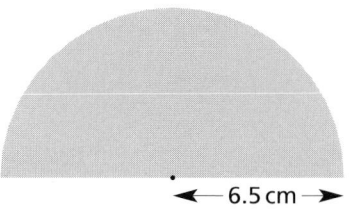

3. Work out the size of angles a, b and c.

   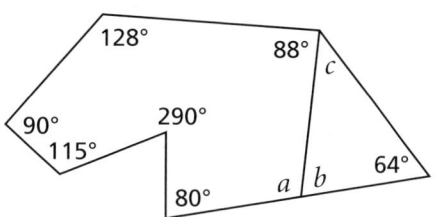

4. Meganair allows passengers to take 20 kg of baggage free.
The table shows the cost (£C) for some weights of excess baggage (W kg).

W	5	10	12	15
C	12.50	25.00	30.00	37.50

   (a) Work out the ratio $\frac{C}{W}$ for each pair of values.

   (b) Is C directly proportional to W?

   (c) Find an equation connecting C and W.

   (d) Use your equation to find the cost of 13 kg of excess baggage.

5. Calculate the radius of a circle with area 200 cm², correct to 1 d.p.

6 Find the missing angle.

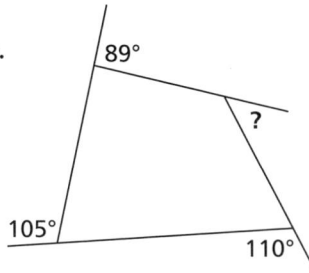

7 Kim makes lilac gold for a ring by mixing 12.6 grams of pure gold with 4.2 grams zinc.

To make lilac gold for a brooch, how many grams of zinc would she need to mix with 24 grams of pure gold?

8 This sequence of toucans continues to the right.

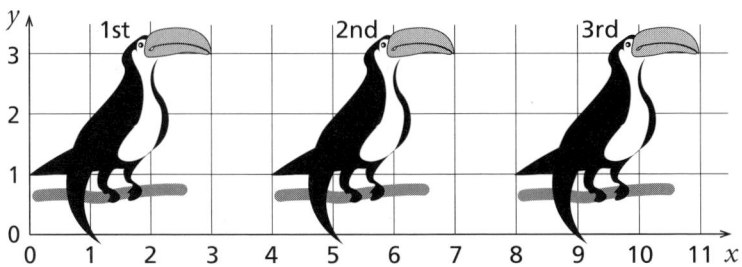

(a) What are the coordinates of the 1st toucan's beak tip?

(b) Show coordinates for the first six toucans in a table like this.

Beak coordinates		
Toucan (n)	x	y
1	3	3
2		

(c) What are the coordinates of the nth toucan's beak tip?

(d) Work out the coordinates of the 50th toucan's beak tip.

9 This diagram shows a regular pentagon and a regular nonagon (9-sided polygon).

Work out the size of angle x.

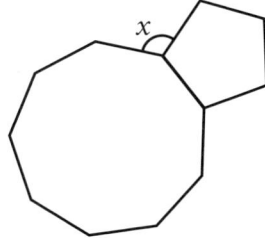

10 A regular hexagon is drawn inside a circle.
Every vertex touches the circumference.
The diameter of the circle is 10 cm.

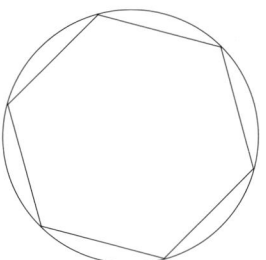

(a) Work out the area of the whole circle, to 1 decimal place.
(b) What is the length of one edge of the hexagon?
(c) A formula that gives the approximate area of a regular hexagon is
$$A = 2.6s^2$$
where A is the area and s is the length of one edge.
Use the formula to work out the approximate area of the hexagon.
(d) What percentage of the area of the whole circle is the area of the hexagon inside it?

11 In the table, p is directly proportional to q.
Find the missing value.

q	3.6	7.8	10.2
p	12.6	■	35.7

12 Work out the exterior and interior angles of an 18-sided regular polygon.

Do not use a calculator for questions 13 to 17.

13 Decide whether each of these is true or false.
(a) $(0.5)^2 = 0.25$
(b) $\sqrt{0.04} = 0.02$
(c) $\sqrt{0.64} = 0.8$

14 What is the cost of 6 pizzas at £6.79 each?

15 Find the nth term of this linear sequence.
1.2, 1.5, 1.8, 2.1, 2.4, …

16 Add the lengths 3.21 km and 560 metres.

17 Work these out.
(a) 3.2×0.24
(b) $8 \div 0.2$
(c) $0.84 \div 1.2$
(d) $7.28 \div 1.3$

19 The right connections

Section B

Here is some information on some species of whales.

Species	Mean length (metres)	Mean weight (tonnes)	Cruising speed (knots)	Number of vertebrae
Great Right	15.0	96.0	5	57
Bowhead	16.0	110.0	3	54
Grey	12.2	34.0	6	56
Fin	21.0	70.0	20	62
Blue	25.0	178.0	13	64
Minke	8.2	9.0	15	48
Sei	17.0	29.0	26	57
Humpback	14.6	48.0	4	53
Goosebeak	6.4	4.5	3	47
White	4.3	1.3	3	51
Great Sperm	15.0	38.0	3	40
Great Killer	8.0	6.0	5	53

1. (a) Draw a scatter diagram for the length and weight of each species.
   (b) Describe any connection between the lengths and weights of these whales.

2. (a) Which whales are not very heavy compared to their lengths?
   (b) What are the cruising speeds of these whales?
   (c) What might be the reason why these whales are able to go at these speeds?

Sections C and D

1. A marine biologist suggests that heavier whales go slower.
   She draws a scatter diagram from the data on the previous page to show this.

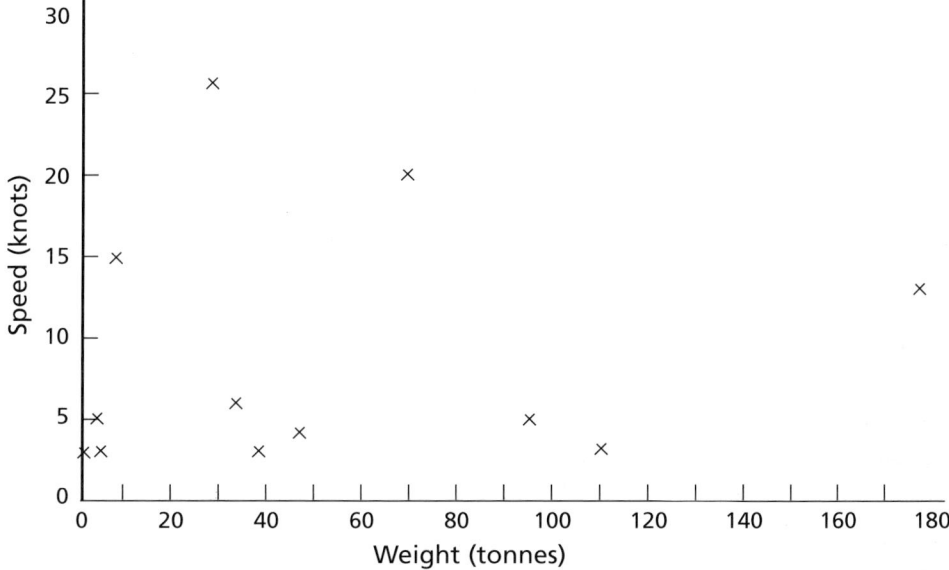

   (a) Describe the correlation between weight and cruising speed.
   (b) Do you agree with the marine biologist's hypothesis?

2. (a) Draw a scatter diagram of the length and the number of vertebrae.
   (b) Draw a line of best fit on your scatter diagram.
   (c) Use your line to estimate the number of vertebrae you would expect to find in a species of whale for which the mean length is 22 m.

3. Use the quartering method to examine the correlation between length and number of vertebrae.

20 Algebra problems

Section B

1 (a) Copy and complete this working to solve $35 - 4n = 6n - 5$.

$35 - 4n = 6n - 5$ (add $4n$ to both sides)
$35 = \ldots - 5$ (add 5 to both sides)
$40 = \ldots$ (divide both sides by 10)
$\ldots = \ldots$

   (b) Check that your answer works in the original equation.

2 (a) Copy and complete this working to solve $25 + 2x = 95 - 5x$.

$25 + 2x = 95 - 5x$ (add $5x$ to both sides)
$ = $ (take 25 off both sides)
$ = $ (divide both sides by 7)
$ = $

   (b) Check that your answer works in the original equation.

3 This equation has been solved incorrectly.
   Write out the correct solution.

$20 - 4n = 5 + n$ (take $4n$ from both sides)
$20 = 5 - 3n$ (take 5 from both sides)
$15 = 3n$ (divide both sides by 3)
$5 = n$

4 Solve each of these equations.
   (a) $5g + 12 = 92 - 3g$
   (b) $50 - 2j = 5j - 6$
   (c) $41 - 2y = 77 - 5y$
   (d) $100 - 4r = 76$
   (e) $30 - k = k - 22$
   (f) $35 - 2h = h + 8$
   (g) $74 - 5e = 39 + 2e$
   (h) $100 - 7q = 61 - 4q$

Section C

1. Solve each of these number puzzles.

   (a) I think of a number.
   I multiply it by 6.
   I take my result away from 100.
   The answer is 4 times the number I started with.
   What number was I thinking of?

   (b) I think of a number.
   I multiply it by 4.
   I take my result away from 56.
   I get the same answer if I multiply my starting number by 7 and take the result away from 92.
   What number did I start with?

2. Lee and Alice are playing number puzzles.

   I multiply my number by 6 and then take 24 away from the result.

   I double my number, then take the result away from 40.

   They both start with the same number, and their answers are the same.

   What number did they start with?

3. Copy and complete the solution of this equation.

   $4(5 - y) = 3(y + 2)$ (multiply out brackets)
   $20 - 4y = 3y + 6$ (add $4y$ to both sides)
   $20 = \ldots + 6$ (take ... from both sides)
   $\ldots = \ldots$ (divide both sides by ...)
   $\ldots = \ldots$

4. Solve each of these equations.

   (a) $2(7 - x) = 3x - 1$
   (b) $2(12 - 3a) = 5(4 - a)$
   (c) $3(10 - r) = 4(15 - 2r)$
   (d) $2(v - 2) = 5(9 - v)$
   *(e) $10 - 3b = 9 - 4b$
   *(f) $5(1 - w) = 2(4 - 2w)$

Section D

*1 Greg and Emma are decorating cakes.
They put the same number of
chocolate drops on each cake.
Greg starts with 95 drops and
Emma starts with 107.

When they finish Greg has decorated 15 cakes
and Emma has decorated 17 cakes.
They both have the same number of chocolate drops left over.

Suppose each cake uses n chocolate drops.

(a) Write an expression using n, for the number of drops
Greg has left over.

(b) Write an expression using n, for the number of drops
Emma has left.

(c) Use your answers to (a) and (b) to form an equation.
Solve it to find n.

(d) How many chocolate drops did they use to decorate each cake?

(e) How many drops do they each have left?

*2 Fraser and Naomi are writing their Christmas cards.

Fraser starts with 6 packs of cards.   Naomi starts with 5 packs of cards.

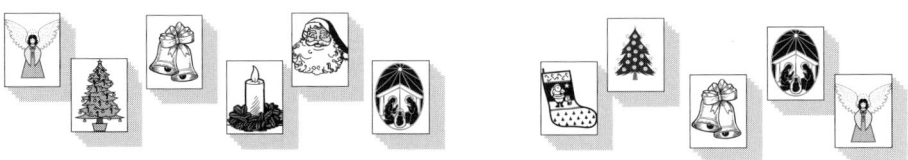

Each pack contains the same number of cards.

Fraser writes 25 cards and Naomi writes 19 cards.
Now they each have the same number of cards left.

(a) How many cards are in a single pack?

(b) How many cards do they each have left?

*3 John and Amjid work in a restaurant and earn the same hourly rate.
One evening John works for 4 hours and also earns £5.50 in his tips.
Amjid works for 3 hours and earns £10 in tips.
They both earn the same total amount that evening.

(a) How much is their hourly rate?

(b) How much did they each earn that evening?

*4 Look at this set of patterns.
Each one is made from matchsticks.

Pattern 1   Pattern 2   Pattern 3

   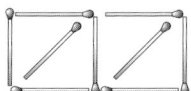   

(a) Make a table going up to pattern number 5,
showing how many matchsticks there are in each pattern.

Pattern number	1	2	3
Number of matchsticks	5		

(b) Copy and complete this formula.

Number of matchsticks in the nth pattern = ... n + ...

(c) In one pattern there are 1449 matchsticks.
Using your answer to (b) write down an equation for n.

Solve your equation to find the pattern number
which has 1449 matchsticks.

*5 Solve each of these equations.
Some of them have more than one answer.

$\dfrac{3}{a} = \dfrac{1}{7}$

$\dfrac{b^2}{3} = 27$

$\dfrac{2}{c} = \dfrac{4}{10}$

$d^2 + 5 = 30$

$4e^2 = 100$

$\dfrac{3f-2}{7} = 1$

$\dfrac{5g-7}{3} = 1$

$\dfrac{1}{3-4h} = 1$

$\dfrac{10}{3-j} = 2$

21 Angles

Section A

1 Find the missing angles.

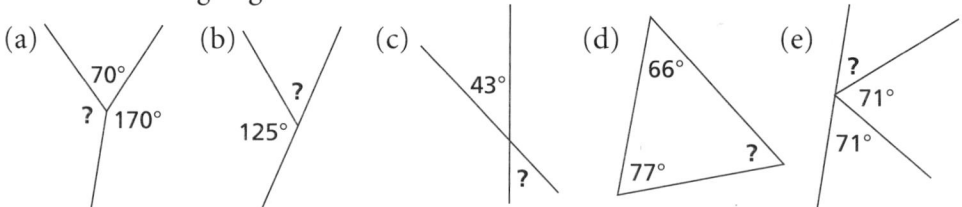

2 Find the missing angles, writing brief justifications for your results.

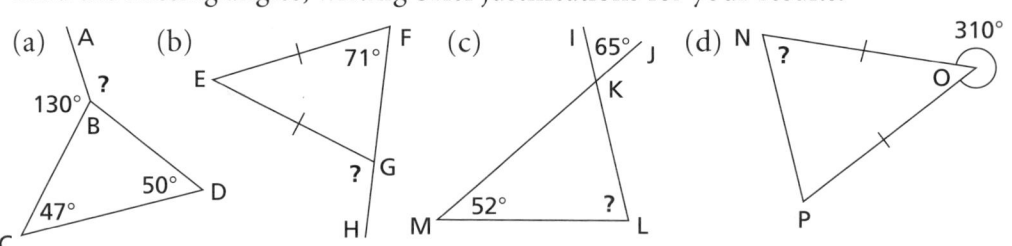

Section B

1 Pairs of equal angles are marked here.
   What is each type called?

   (a)   (b)   (c)

2 Find the missing angles, writing brief justifications for your results.

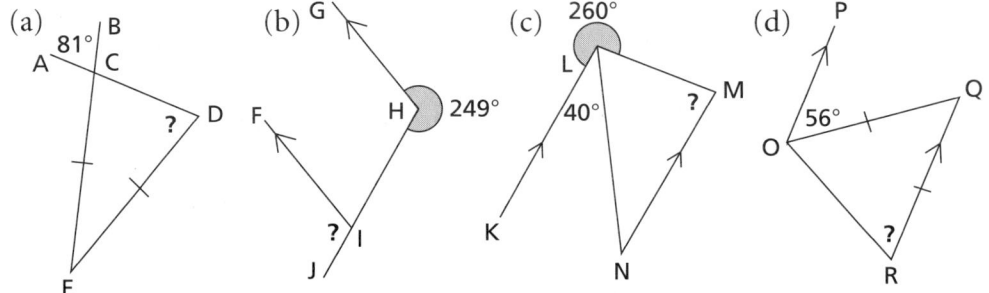

22 Transformations

Sections A to D

1. A sailing dingy tacks down the lake from **A** towards **B**.

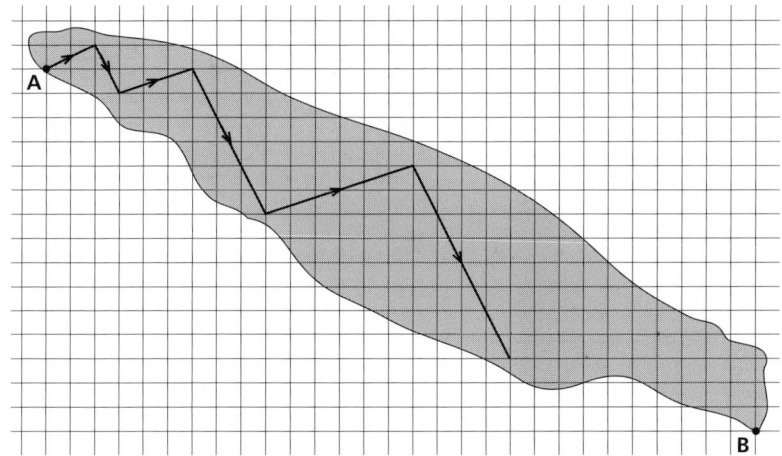

The first three tacks of its voyage are described by these column vectors. $\begin{bmatrix} 2 \\ 1 \end{bmatrix}, \begin{bmatrix} 1 \\ -2 \end{bmatrix}, \begin{bmatrix} 3 \\ 1 \end{bmatrix}$

(a) Write column vectors for the next three tacks.

(b) The boat needs to make two more tacks to reach B. What might they be?

2. (a) Make a copy of triangle ABC.

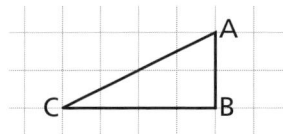

Reflect the triangle in the line AB and draw the image.
What is the name of the shape formed by triangle ABC and its image?

(b) Make another copy of triangle ABC.
Mark the point half-way along the line AB and label it X.

Draw the image of the triangle after a 180° rotation about X.
What is the name of the shape formed by the triangle and its image?

The repeating pattern below is formed from congruent triangles.
Some shapes have been labelled and some points and lines are shown to help describe any transformations.

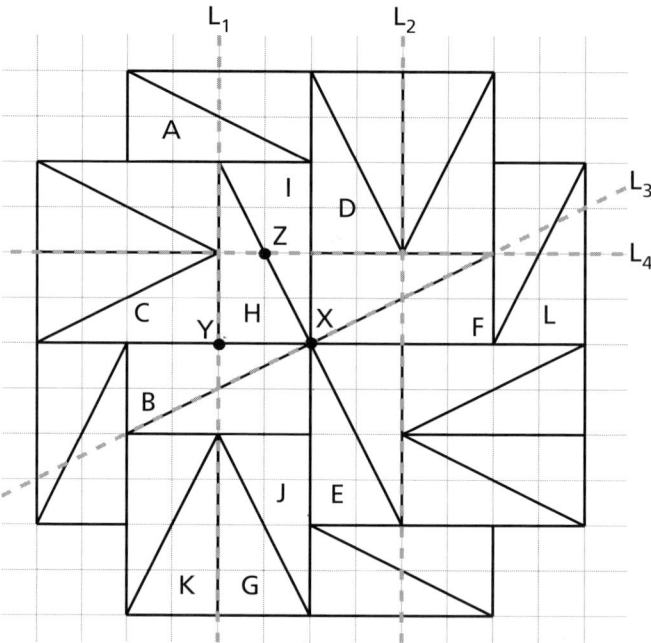

3 Which shape is the image of
  (a) shape K after the translation $\begin{bmatrix} 8 \\ 6 \end{bmatrix}$
  (b) shape H after reflection in the line L_2
  (c) shape J after a rotation of 180° about point X
  (d) shape B after an anticlockwise rotation of 90° about X
  (e) shape D after the translation $\begin{bmatrix} -2 \\ -8 \end{bmatrix}$
  (f) shape C after a clockwise rotation of 90° about X

4 Describe fully a transformation that maps
  (a) H to E (b) G to K (c) I to H (d) C to H (e) E to I
  (f) A to B (g) I to J (h) B to I (i) D to C (j) J to C

5 Draw a grid like this with each axis numbered from ⁻8 to 8.

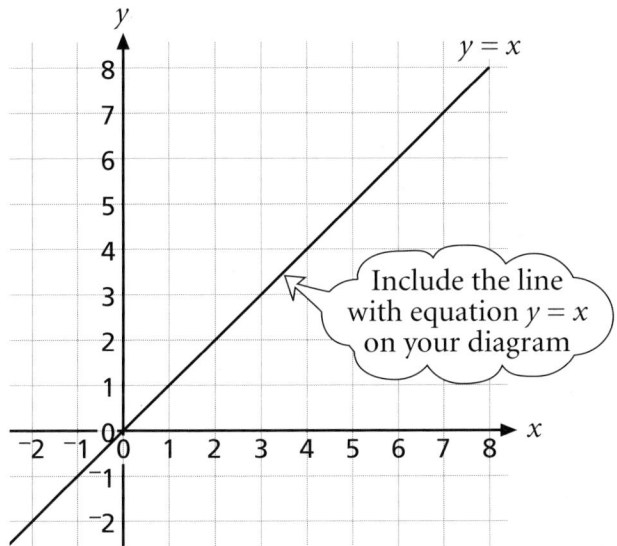

Include the line with equation $y = x$ on your diagram

(a) Plot the points (3, 3), (8, 3), (5, 1) and (3, 1).
Join them up and label the shape A.

(b) (i) Reflect shape A in the y-axis.
Label the image B.

(ii) Reflect shape A in the x-axis.
Label the image C.

(iii) Reflect shape A in the line $y = x$.
Label the image D.

(iv) Rotate shape A anticlockwise 90° about the point (0, 0).
Label the image E.

(c) (i) What transformation will map shape E onto D?

(ii) What transformation will map shape B onto C?

(iii) What transformation will map shape D onto C?

Section E

1. What single translation would have the same effect as translating by $\begin{bmatrix} 6 \\ 3 \end{bmatrix}$ then $\begin{bmatrix} 5 \\ -2 \end{bmatrix}$?

2. Translating by $\begin{bmatrix} a \\ 7 \end{bmatrix}$ then $\begin{bmatrix} 3 \\ b \end{bmatrix}$ has the same effect as translating by $\begin{bmatrix} -2 \\ 9 \end{bmatrix}$. Find the values of a and b.

3. Copy the shape and lines A and B on to the middle of some squared paper.

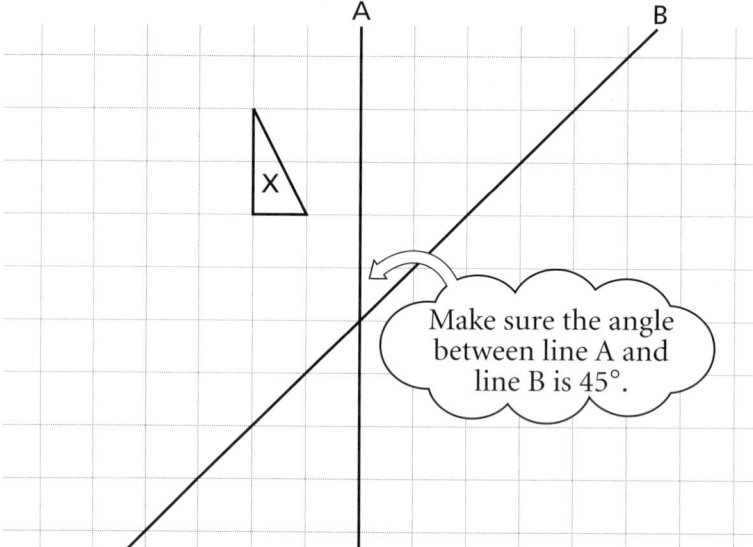

Make sure the angle between line A and line B is 45°.

(a) (i) Transform shape X by a reflection in line A followed by a reflection in line B. Label the final image Y.

(ii) What single transformation could you use to map X on to Y?

(b) Now draw your own shape on the diagram.
Reflect it in line A and then line B.

Try this with different shapes.

What do you notice?

Some congruent L-shapes are shown on the diagram below.

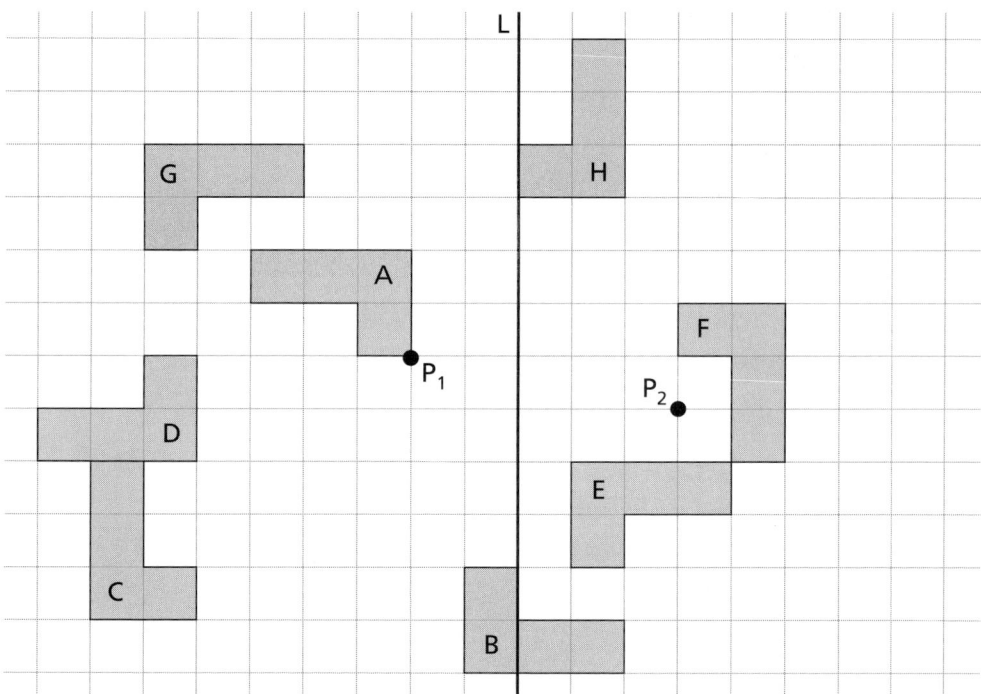

4 (a) What is the image of D after a rotation of 180° about the point P_1 followed by a translation of $\begin{bmatrix} -9 \\ 2 \end{bmatrix}$?

(b) What is the image of E after a reflection in line L followed by a rotation of 90° clockwise about the point P_2?

5 Use these transformations in this question.

| Rotate 180° about the point P_1 | Translate by $\begin{bmatrix} 1 \\ -4 \end{bmatrix}$ |

| Rotate 90° anticlockwise about the point P_2 | Reflect in line L |

What **pair** of these transformations (and in what order) will map
(a) A to B   (b) A to E   (c) B to C   (d) C to F

23 Trial and improvement

Sections A and B

For each question show your trials clearly.

1. Two consecutive numbers multiply to give 2162.
   Sean tries
   $$41 \times 42 = 1722 \text{ too small}$$
   Find the two numbers.

2. Find three consecutive numbers that multiply together to give 79 464.

3. The length of this rectangle is 7 cm more than its width.
   Its area is 494 cm².
   How long and how wide is it?

4. $n(n + 1) = 6162$
   Find n.

5. Two numbers differ by 4 and multiply to give 957.
   Find the numbers.

6. Solve the equation $n^3 = 9.261$.

7. Solve the equation $n^2 - n = 4.16$ where n is a positive number.

8. A rectangle has an area of 64.8 cm².
   Its length is 5 times its width.

   Copy and complete the table below to find the width and length of this rectangle.

Width (cm)	Length (cm) (width × 5)	Area (cm²)	Result Too big Too small
2	10	20	

Section C

1. The length of a rectangle is 3 cm more than its width. Its area is 36 cm².

   width + 3 / width

   Find the width and length of this rectangle, correct to 1 decimal place. Start by trying a width of 5 cm.

2. The length of a rectangle is three times its width. Its area is 60 cm².

   Find the width of this rectangle, correct to 1 decimal place. Start by trying a width of 4 cm.

3. Ross is solving the equation $n^3 = 37$ by trial and improvement.
   He tries $3 \times 3 \times 3 = 27$ too small
   $3.5 \times 3.5 \times 3.5 = ...$

   Solve the equation correct to 1 d.p.

4. Find a solution of the equation $x^2 - x = 15$, correct to 1 d.p. Start by trying $x = 4$.

5. Find the solution of the equation $x^3 + x = 200$, correct to 1 d.p.

6. S - Can you make me a fish tank?
   The height should be 10 cm more than the width, and the length must be twice the height.
   Capacity to be 70 000 cm³.
   Ta - Jim

   Work out the dimensions of Jim's fish tank correct to 1 d.p.
   You could use a table like this to record your results.

   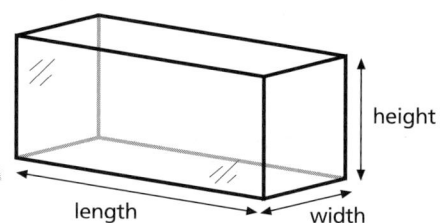

Width (cm)	Height (cm)	Length (cm)	Capacity (cm³)	Too big/ Too small
20	30	60		

Mixed questions 4

1. Jo wants to know if fuel economy is related to engine capacity for cars. She collects some data on 18 cars at random.
   Here are her findings.

Engine capacity (litres)	Fuel economy (max. miles per gallon)
1.8	47
1.4	50
2.5	44
1.6	52
1.0	55
1.6	50
1.8	56
2.3	41
1.4	57
1.4	53
2.5	38
1.6	46
1.2	51
2.0	38
0.9	60
1.6	55
3.0	38
0.9	53

   (a) Draw a scatter diagram for the engine capacity and fuel economy of these cars.

   (b) Describe any connection this data shows between engine capacity and fuel economy.

   (c) Use the quartering method to examine the correlation between engine capacity and fuel economy.

2. Find the missing angles.

   (a)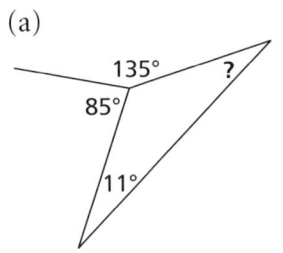

   (b)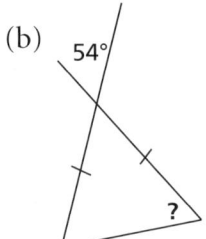

   (c)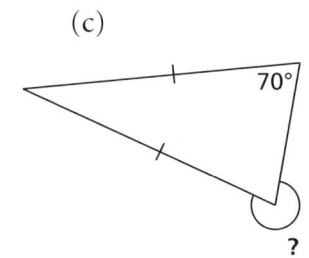

3   Solve these equations.
   (a) $4x - 5 = 5$
   (b) $5p - 1 = p + 19$
   (c) $3m - 2 = 7m - 18$
   (d) $2(n - 1) = 3(n - 3)$

4   This eight-pointed star can be made by joining eight points equally spaced round a circle.

   (a) Describe the rotation symmetry of this star.
   (b) How many right angles are these in this design?
   (c) Give a pair of parallel lines.
   (d) Write down a pair of
      (i) vertically opposite angles
      (ii) alternate angles
      (iii) corresponding angles

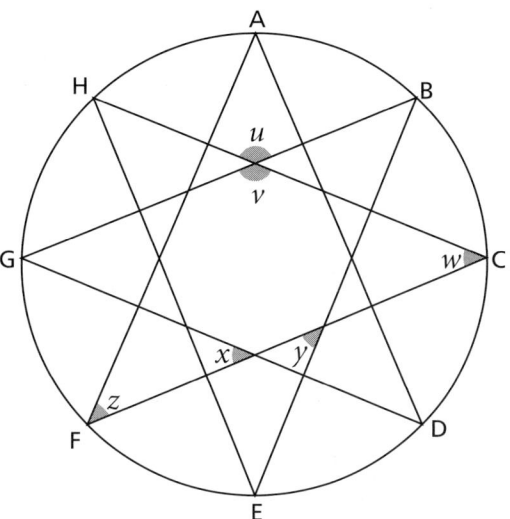

5   Solve these equations.
   (a) $3f + 1 = 23 - 8f$
   (b) $20 - t = 5 + 2t$
   (c) $3 - 2n = 21 - 5n$
   (d) $5y - 14 = 35 - 2y$

6   On squared paper, draw a pair of axes both numbered from ⁻8 to 8.
   (a) Draw the lines with equations $y = 1$ and $x = 3$.
   (b) Plot and join the points $(1, ^-1)$, $(1, ^-4)$ and $(^-1, ^-4)$. Label the shape A.
   (c) Transform shape A using a reflection in the line $x = 3$ followed by a reflection in the line $y = 1$. Label the final image B.
   (d) What single transformation will map shape A onto shape B?
   (e) Now translate shape B by $\begin{bmatrix} -6 \\ -2 \end{bmatrix}$ and label the image C.
   (f) What single transformation will map C onto A?

7 The angles of this triangle, in degrees, are as shown.

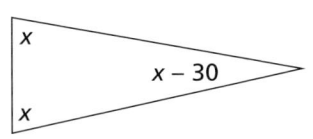

(a) Which of these equations is correct? Explain your choice.

$x - 30 = 180$	$2x - 30 = 90$
$2x - 30 = 180$	$3x - 30 = 180$
$3x - 30 = 90$	$x - 30 = 90$

(b) Solve this equation to find the value of x.

(c) What is the smallest angle of the triangle?

8 Find the missing angles, writing brief descriptions of your method.

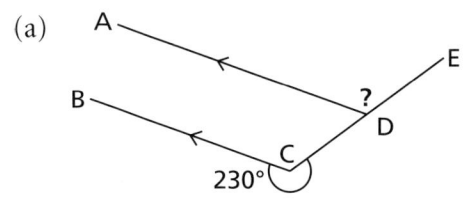

9 Solve each of these number puzzles by writing an equation and solving it.

(a) I think of a number.
I double it and then take the result away from 40.
The answer is three times the number I started with.
What number did I start with?

(b) I think of a number.
I multiply it by 4.
Then I take the result away from 50.
The answer is the number I started with.
What number did I start with?

10 The dimensions of a cuboid are to be as shown. The volume is to be 10 cm^3.

Use trial and improvement to find the width of the cuboid, to 1 d.p.

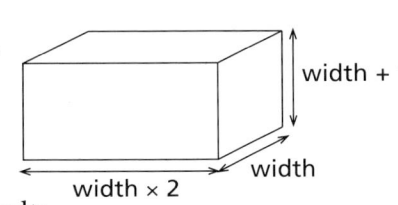

You could use a table like this to record your results.

Width (cm)	Height (cm)	Length (cm)	Volume (cm³)	Too small/ too big